Health & Safety Executive

A report by HM Factory Inspectorate

100 Practical applications of noise reduction methods

WITHDRAWN

LONDON: HER MAJESTY'S STATIONERY OFFICE

ISBN 0 11 883691 9

Foreword

For those who are faced with noise problems in industry the reduction of noise at source by the application of engineering solutions is clearly a primary aim since it generally resolves the problem once and for all. Sometimes however, although anxious to take action, people in industry are unsure about the best way to tackle what may be quite a common problem, or else they find that there are alternative ways of dealing with the matter. This series of case studies is not a comprehensive treatise but is intended to show some solutions which have been adopted by industry to deal with noise problems. The methods illustrated have proved to be effective in practice; the rewards from the expenditure involved have been considered very worthwhile by the companies involved. The intention of the studies is not to give detailed guidance about precisely how a job should be done but to show some of the ways which can be used. Again a principle which has been established for one industry may well have a wider application in other fields, since such cross-fertilisation is often productive.

The HSE will continue to collect details of effective solutions to noise problems and if the volume of information justifies it a further series of case studies may well be forthcoming.

J D G HAMMER
HM Chief Inspector of Factories

UK Scene – Timber moulders

Noise exposure reductions by enclosure of Multi Spindle Moulding machines in the Woodworking Industry have been shown in previous studies in this series. Pages 84, 85, 86, 87, 88, 89, show various types and features of individual enclosures.

It has been stressed that inclusion of any method only means that it was a practicable way of reduction in that particular situation but not necessarily reasonably practicable, in the legal sense, for all other cases. Other considerations could still mean an individual assessment is needed.

The conclusion reached may be influenced by existing widespread acceptance. The information below avoids individual detail to show use throughout the United Kingdom.

Magnet Joinery Ltd, Keighley, West Yorkshire

Palatial Ltd, London E3

Mallinson Denny Ltd, Cardiff

Baynes (Reading) Ltd, Wheatley, Oxon.

George Rankin and Co Ltd, Belfast

Sandell-Perkins Ltd, Aylesford, Kent

J Fleming (Southern) Ltd, Aberdeen

W H Shaw and Son Ltd, Oldham, Greater Manchester

Acknowledgement and thanks to the Northern Ireland Factory Inspectorate and the other contributors for their co-operation.
A single page format is used to indicate acceptance of a particular solution throughout the country.

Introduction

Inspectors have found, in general, a willingness in industry to reduce noise but a lack of knowledge regarding what was possible.

Information covering successful examples of noise reduction – often including photographs, brief notes of the principles used, where expertise can be obtained, an estimate of cost and amount of reduction – have been circulated informally and in reports as a guide towards what can be achieved.

To increase the rate and case of circulation information was gathered together from various sources, prepared in similar single sheet format and distributed within HM Factory Inspectorate.

These have proved extremely useful and popular. It is in response to innumerable requests that some 100 of this series have been brought together in this report to make them available to the general public.

There are a number of methods available to assess objectively the efficiency of an acoustic treatment. Strictly, 'Noise Reduction' means the difference in sound levels, at the same measuring position, before and after treatment. It has been impossible to take such measurements for some of the treatments. In these cases the 'Sound Level Differerence' has been quoted e.g. measurements inside and outside an enclosure. Furthermore, where the term 'claimed' is used the information is reported in good faith but unverified by the Health and Safety Executive.

This series is intended as a guide only and should be used with care since its purpose is not to specify in detail how noice reductions can be carried out but indicate what might be achieved with application and expertise. The Health and Safety Executive does not warrant the accuracy of statements made nor the information contained in the guide nor have any of the methods been given official approval.

INCLUSION DOES NOT MEAN THAT THE METHOD SHOWN WILL BE CONSIDERED REASONABLY PRACTICABLE UNDER HEALTH AND SAFETY AT WORK LEGISLATION.

Each situation will still be subject to assessment.

Contents

Low noise design

Low noise design

Road breaker

A new approach to the design of pneumatic road breakers has considered noise and vibration as major parameters. Heavy forgings have been replaced by steel and castings moulded in polyurethane.

Plastics bonded to internal components adds to acoustic damping. The integrated outer case allows long attenuating exhaust passages giving further noise reductions.

Electrically and thermally insulated handles are additional safety features, and provide significant vibration reduction.

Approximate noise dose at operator ear

110dB(A) Conventional Road Breaker fitted with no muffler and undamped moil point

107dB(A) Conventional Road Breaker fitted with muffler and undamped point

106dB(A) Conventional Road Breaker fitted with muffler and damped moil point

103dB(A) New design fitted with damped moil point

N.B. The noise measurements quoted are for machine comparison purposes only. They DO NOT represent absolute measures by British Standard or other standard methods and the levels MUST NOT be compared with EEC 82.

Noise reduction	**approx 7dB(A)**
Weight reduction	**20lb**
Cost reduction (1981)	**25%**

Acknowledgements to Tilbury Plant Ltd, Maidstone, Kent for cooperation in making these measurements
Further details Compair Construction and Mining Ltd, Cambourne, Cornwall

Rotary swaging

Swaging is a process used to modify the shape of metal tubes and solid bars. Conventional rotary swaging machines operate by hammering the workpiece in a split die box until the workpiece is formed to the same shape as the die. This process is inherently very noisy. A new design of machines* has been developed which carries out the same process with much less noise. The new machine uses hydraulic pressure to force the one piece die box around the workpiece instead of hammering. The new method results not only in reduced noise but also in a considerable saving in material on tube work since a greater proportion of material goes into elongation and not wall thickening as with conventional swaging methods.

Additional advantages include:
1) Will accept greasy material – pre production cleaning unnecessary
2) Workpiece held in a vice – no operator contact with vibrating workpiece
3) Automatic feed, programmable
4) Four station rotary head may be used in a single, two, three or four-die cycle for quick setting

Conventional Machine 98–108dB(A)

Hydraulic Push Pointer 75–80dB(A)

Claimed noise reduction 23–28dB(A)
Additional cost (1981) NIL (at about £31,000 the cost is
roughly equivalent to conventional machines)

* Developed, manufactured and marketed by Stevens and Bullivant Ltd, Spring Hill, Birmingham.

Low noise alternative to rivetting

High strength break-neck fastening systems* have been developed to replace rivetting or conventional nuts and bolts. High noise levels associated with rivetting hammers or pneumatic torque wrenches as well as that generated by the structures being fastened are eliminated. Instead of hammering a heated rivet into position the new break-neck fastener is secured by swaging a collar onto pre-machined annular grooves on the bolt.

Fastening cycle

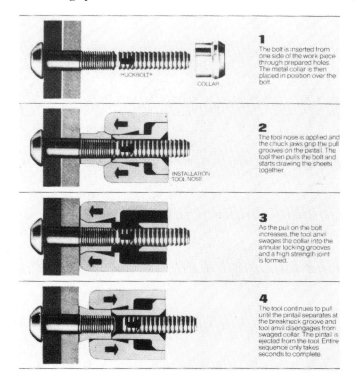

1
The bolt is inserted from one side of the work piece through prepared holes. The metal collar is then placed in position over the bolt.

2
The tool nose is applied and the chuck jaws grip the pull grooves on the pintail. The tool then pulls the bolt and starts drawing the sheets together.

3
As the pull on the bolt increases, the tool anvil swages the collar into the annular locking grooves and a high strength joint is formed.

4
The tool continues to pull until the pintail separates at the breakneck groove and tool anvil disengages from swaged collar. The pintail is ejected from the tool. Entire sequence only takes seconds to complete.

Application of Huckbolt® fastener
Photograph and measurements by permission of British Railways Engineering Ltd, Shildon, Co Durham

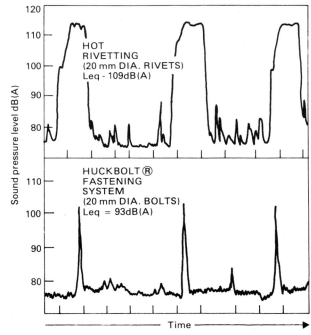

A comparison in noise levels of a break-neck fastening system and hot rivetting.

Additional advantages claimed

1. Component parts locked with a high uniform predictable clamping force.
2. Joints highly resistant to vibrating loosening.
3. Fast effective fastening by semi-skilled operators.
4. Requires only visual inspection.
5. Tension pre-load helps prevent water seepage and corrosion.
6. Easily removed by cutting the collar.
7. Work surfaces not damaged during fastening or removal.
8. No rivet heating furnace required and no danger from handling hot rivets.
9. Reduced exposure of operators to vibration.

Approximate noise reduction: 15–20dB(A)
Estimated Cost: Increased cost of fasteners may be offset against quicker fitting, reduced labour and the advantages listed above.

* Avdelok System (range ³⁄₁₆″–³⁄₈″ dia) supplied by Avdel Fasteners Ltd, Welwyn Garden City, Herts.
Huckbolt® System (range ³⁄₁₆″–1⅛″ dia) supplied by Thomas William Lench Ltd, Warley, West Midlands.

Skid rails

Lamberton & Co Ltd have redesigned the basic skid rail used in steelworks reducing the noise generated by the 'slip stick' action when conveying steel joists by this traditional manner.

The new rail consists of 2 staggered rows of free running discs mounted with their vertical axis slightly skewed.

Patents apply to this new design which is being marketed by the company.

It is claimed that noise levels, often between 115 and 120dB(A), can be reduced by up to 35dB(A) resulting in 80 to 85dB(A).

An additional advantage is that this method gives approximately 25% saving in energy.

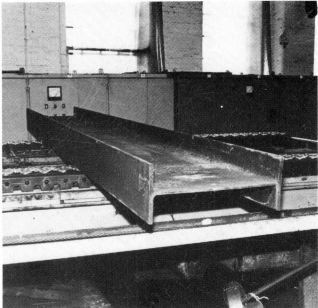

Claimed noise reduction 30–35dB(A)
Cost (1980) in the order of £150 per foot length excluding site modifications

Supplied by Lamberton & Co Ltd, Coatbridge, Lanarks

New design reduces noise of bowlfeeders

A new design of bowlfeeder has been developed* which operates on a different principle from conventional bowlfeeders but may be retrofitted. The new design is much quieter than conventionally operating machines and incorporates a unique load sensing system which replaces the conventional paddle. Both horizontal and vertical motions of the bowl are individually sprung and separately powered. Separate controls for these orthogonal motions permit tuning for widely different weights of components without the need to vary bowl springing. A feed rate control varies the firing of the vertical drive pulse relative to the horizontal pulse allowing the vibration angle to be optimised for the required feed conditions.

The Salford bowl feeder

Advantages over conventionally operating bowlfeeders

1) Low noise level (without acoustic enclosure)
2) Faster feed rate (2 : 1)
3) Smooth feed
4) Low power consumption
5) Low voltage
6) Can feed components other bowls cannot
7) A drive conversion kit is available for conventionally operating bowlfeeder

Noise Levels (21″ diameter bowl)

Component	Feed rate (components/min)	Noise level at 1m from rim dB(A)
Spark plug shells	60	79
UK 10p coin blanks	300	78
Empty bowl	Max. feed rate	57

Claimed noise reduction 10–15dB(A)
Cost £1,500 for 21″ untooled diameter bowl

* Developed and marketed by Salford University Industrial Centre Ltd, Salford

Laminar flow fans

A hydraulic power pack was to be used to drive a test rig. Cooling for this power pack was effected by two standard 16″ diameter axial 6-blade fans inducing air through the top and out of a side mounted radiator. With the running at 3000 rpm the prototype unit produced an unacceptable 98dB(A) pwl.

Consultant specialists recognised this as a suitable application for Laminar Flow Fans. Possibly the first commercial use of low noise air moving technology still being developed. These replacement fans are of approximately 77dB(A) pwl each (twin fans producing 80dB(A)).

This level is not achieved by the total power pack because other sources now predominate, but nonetheless the change of fans would enable the rig to meet the customer's specification of no more than 65dB(A) spl at 4m.

Power pack with covers and inlet duct removed to show fans

Axial replaced by laminar fan

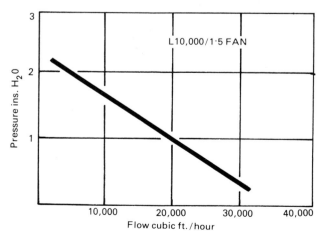

Performance

Photographs by permission of Twyford Moors Aircraft and Engineering Ltd, Eastleigh, Hants

Claimed reduction 18dB(A)
Estimated on-cost (1982) £100

Fan design by Wolfson Unit for Noise and Vibration control ISVR, University of Southampton. Manufactured by Twyford Moors under licence from British Gas Corporation (UK Patent No 1461776)

Double end tennoning machine

The need to reduce levels of hazardous noise is being recognised by manufacturers who are incorporating noise control features into basic designs.

Double end tennoning has been recognised as a particularly bad source of noise especially when cutting plastic laminates.

The machine shown above incorporates anti-vibration mountings to isolate internal sources and an enclosing acoustic guard as an integral part of the machine allowing cutters to be located remote from necessary access openings.

The resulting sample Leq in this particular installation was 85dB(A) which means that the noise requirement under Section 44 of the Woodworking Machines Regulations 1974, has been satisfied without the use of ear protection.

It is difficult to estimate the additional cost of treatment specifically included for noise control since so many of the features are necessary anyway, simply being included design, while other reductions come from features giving other advantages for instance the cutter head braking which increases safety.

Photograph by permission of Stephen Laminates, Glenrothes, Fife

Noise reduction 15–20dB(A)
Estimated additional cost (1981) 5% – see note above

Semi-automatic core knockout in a foundry

The traditional method of removing sand or ceramic moulding cores from castings is to use a hand-held pneumatic percussive hammer, as seen at one factory producing aluminium components, (114dB(A) sample Leq). The operator needs to wear ear and eye protection as well as coping with dust nuisance.

Therefore in order to improve the working environment, this firm has introduced some semi-automatic machines to perform the same task. The operation of the machines is by a variable rate percussion hammer which is clamped to the casting. The hammer clamp carriage is supported on anti-vibration mounts and mounted in a sound reducing cabinet, as shown below. A machine and cabinet in good order were found to produce levels of only 85dB(A) sample Leq beside the operator.

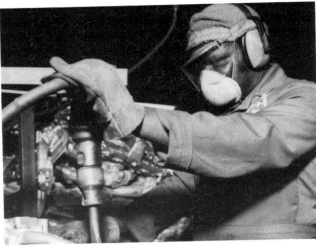

Traditional method

Cabinet shut

Cabinet open

Photographs by permission of Barton Aluminium Foundries Ltd, Birmingham

Other advantages

1. No need to heat treat the cores prior to knockout.
2. Dust extraction facilities can be built in.
3. Enclosure gives protection against flying debris.
4. Vibration insulation.
5. High rate of productivity.

Claimed noise reduction 30dB(A)
Cost (1982) £9,500 (basic machine)

* Machines produced by Epic Engineering Group, Slough, Berks.

Noise and vibration reduction of textile fibrillation plant

The Company manufactures man-made textiles, and one special product involves the fibrillation of woven fabrics.

The machine required for this process was manufactured to the Company's specification and noise control was taken into consideration, as it was known that high levels of low frquency vibration were likely to be generated. Consequently, the machine was fitted with hydraulic anti-vibration mounts set on an isolated concrete sub-floor. In addition, a purpose-made, total enclosure was fitted around the machine, and the resultant reduction in noise levels was 18dB(A).

Photograph by permission of Bonar Textiles Ltd, Dundee

Noise reduction 18dB(A)

Retrospective
treatments
Acoustic guards
Damping
Pneumatics
Room absorption
Barriers and refuges
Enclosures

New support bearings to reduce noise of pipe spinning

A firm manufactures glass reinforced polyester pipe by spinning it inside a rapidly rotating drum, called a mould bed. The original support bearings to the drum created a severe noise hazard which was greatly reduced by using Cooper split plummer block bearings* as shown. Noise reduction was achieved chiefly because:

(i) the bearing cartridges readily swivel to facilitate shaft alignment;

(ii) the new single support wheels are continuously in contact with the mould bed – this was difficult to achieve with the old double-wheel system;

(iii) the load on each support wheel is divided on the rotating shaft by a pair of bearings.

Pipe being withdrawn from mould bed
Photograph by permission of Johnstone Pipes Ltd, Doseley, Telford, Salop

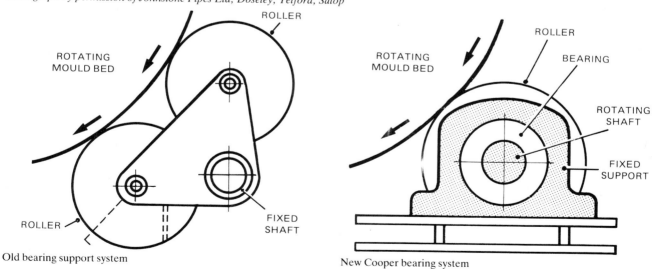

Old bearing support system

New Cooper bearing system

Other advantages: a much simplified layout and hence easy to maintain
Estimated noise reduction 10dB(A)
Cost (1981) £4,000 per mould bed

* Cooper Roller Bearings Co Ltd, King's Lynn, Norfolk

Bronze gears reduce noise

Often noise from intermeshing gears represents a significant contribution to the total noise generated by a machine. Frequent maintenance and adequate lubrication should help to keep such noise to a minimum, however it may still predominate the total noise from a machine.

One company* concerned about the noise from a Goebels Reeler successfully reduced the noise level generated by changing the gearing from steel to bronze. The machine is used for reeling paper of various lengths, it runs at 200–1000 rpm powered by a 20 hp electric motor.

The noise levels measured at the operator position when the machine was fitted with conventional steel gears was 99dB(A) Leq. It appeared that the noise was mainly from the gearing, and so the gears were changed from steel to bronze. After the gear change the noise level was reduced to 86dB(A) Leq.

After running the machine for several years check measurements revealed an increase in noise level to 91dB(A) Leq. The rise in noise was found to be due to poor lubrication. This was rectified and levels reverted to the mid-eighties as previous thus highlighting the importance of adequate lubrication.

Goebels reeler

Bronze gear as fitted for noise reduction

Approximate noise reduction 8–13dB(A)
Cost (1977) £100

* Aarque Systems Ltd, Colnbrook, Berks.

14

Low noise piling

Piling is one of the more noisy tasks in the construction industry capable of generating levels 126–147dB(A). Several specialist piling contractors have already developed quieter methods of installing driven piles. A technique used by one contractor* is to enclose the entire pile and hammer assembly in a vertical acoustic 'box'. The enclosure is of laminated sound proofing construction. The forward face of the box comprises a full-length opening door, through which the pile is passed into the rig. Two smaller doors at the bottom of the enclosure are used to monitor driving progress. The rig box, largest type, is provided with an efficient ventilation system supply for operation. The system can be adapted to accommodate diesel hammers of up to 90 000ft lb energy output and the design can accept most types of pre-fabricated pile. The system may be used in any type of ground and has an additional advantage in sheet piling in that no framework is required. Two sheet piles may be supported in the 'box' at any one time interlocking being made at the ground level. Another feature is safety. After the pile is pitched into the rig, there is no possibility of the pile falling. Even in high winds when traditional piling methods are stopped the Hush rig may continue.

Showing full length opening door

The noise measurements† shown below were taken during a demonstration of an enclosed 3 ton drop hammer lifted by a crane. To enable a comparison to be made with more conventional piling equipment a Compressed Air Driven Hammer 600N was also brought on site.

These measurements were made during piling into 3 metres top layer of sandy clay overlying stiff boulder clay. The rates of penetration were not measured but it was said that the 'Hush Piling' rig was driving two piles into the ground at a faster rate than the air hammer was driving a single pile. Due to the very hard driving and the higher energy output of the drop hammer it must be appreciated that the vibration from the drop hammer was greater than the air hammer although vibration measurements were not made.

Equipment	Distance from source	Impulse sound pressure level dB(A)	Leq (1 min) dB(A)
Compressed air driven hammer (600N)	15m	108	100–103
Hush drop hammer (3 ton)	15m	74–79	63–67

Hush piling in operation

Noise reduction 33–40dB(A) (compared with conventional air hammer).
Additional cost (1982) 2·3% on normal contracts, offset by reduced labour costs (see above).

* SP Civil Engineering Ltd, Brentwood, Essex.
† Information by kind permission Halton Borough Council, Chester

High speed cone winding machines

Noise levels produced by high speed cone winding machines used in the textile industry have been significantly reduced by replacing metal traverse guides with components manufactured from nylon and similar materials. The arrangement of the traverse guides and further details are shown below. A sheet metal guard which normally covers the cams has been removed to reveal the traverse guide drive mechanism.

The traverse guides* have been modified in two different ways; the reduction of noise achieved with each method and their cost differ significantly.

The maximum noise reduction (shown in Fig 3) was achieved by providing a traverse guide manufactured from nylon except for a ceramic insert to improve wear and resistance where the yarn passes through the guide, in combination with a polyethylene tyred cam follower.

A lesser magnitude of noise reduction was achieved at reduced cost by fitting this cam follower to a standard metal guide.

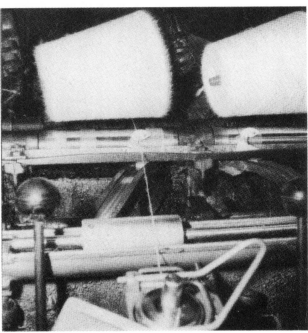

Arrangement of traverse guide
Photographs by permission of Joseph Horsfall, Halifax

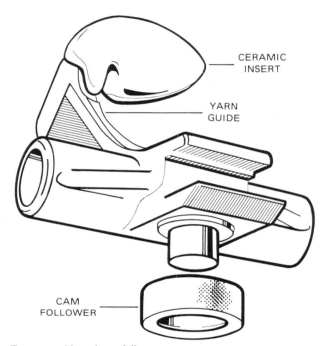

CERAMIC INSERT

YARN GUIDE

CAM FOLLOWER

Traverse guide and cam follower

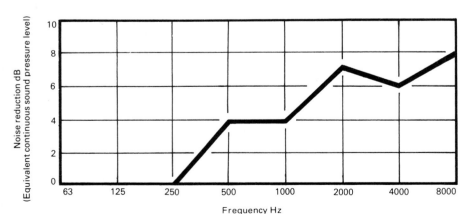

Reduction in octave sound pressure level for nylon traverse guide and cam follower

Approximate noise reduction 7dB(A) for nylon yarn guide and cam follower.
3dB(A) for cam follower with metal traverse guide.
Approximate cost (1982) £4·40/spindle for nylon traverse guide,
£1·40/spindle for cam follower.

* Developed and supplied by Wira, Leeds

Plasma arc cutting

Although plasma arc cutting has been in use for over 20 years, the advent of water injection some 10 years ago gave greatly improved quality of cut and has resulted in a rapid adoption of the plasma arc process for the profile cutting of plate.

It is unfortunate that the benefits gained in production have been associated with exceedingly high noise levels, typically 110dB(A) at 2–3 metres from the arc.

A system is now available* for plasma cutting under water and the reduction in noise level is dramatic. This benefit has been gained with no loss in cutting speed.

Other advantages claimed for underwater plasma cutting are:
  (i) A reduction in the level of fume produced may obviate the need for additional fume extraction.
 (ii) A reduction in ultra-violet radiation from the arc.
(iii) A reduction in thermal distortion of the workpiece.

A typical installation

Assuming that suitable control systems are already in use (particularly nozzle height control) the main additional cost involved is the provision of a water table, typically £7,000 for a 5m × 2m table. The capital cost of the associated cutting machine would be £40 000–£100 000 depending on size of machine.

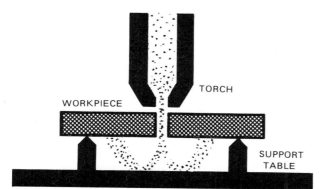

Conventional plasma arc cutting

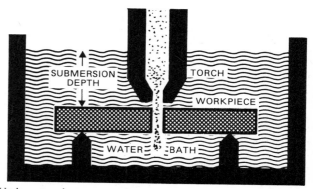

Underwater plasma arc cutting

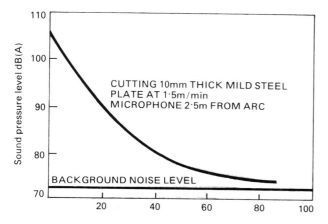

CUTTING 10mm THICK MILD STEEL PLATE AT 1·5m/min MICROPHONE 2·5m FROM ARC

BACKGROUND NOISE LEVEL

Increasing depth of submersion of workpiece (mm).

Effect of submersion of workpiece in water on noise levels

Claimed noise reduction 30dB(A)
Approximate cost (1981) – additional £7,000 over conventional system

* System marketed by Messer Griesheim Ltd, Seaton Delaval, Tyne and Wear

Noise reduction by redesign of power press tooling

The technique of extending the work range of a power press by including a shear or skew cut in blanking tools has been known for some time. A reduction in noise levels has also resulted from this technique. Distortion of the product shape, e.g. ellipticity in circular products, is a major difficulty when tools with simple shear surfaces are employed.

Noise reductions on impact of between 8 to 10dB(A) are claimed by the firm which developed the die shape shown below. The product in this case is a large circular stamping 400mm (16 in) diameter from 2mm (0.8 in) thick aluminium sheet.

In descending the flat face punch first contacts, then grips and finally punches through the plate at the projecting central lands, followed by a progressive symmetrical shear to complete the blank. There was no noticeable distortion in the flatness of the punched plate and it is understood that circularity can be kept within :003" (:075mm) tolerance.

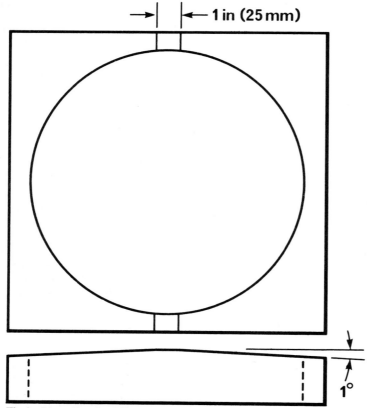

← 1 in (25 mm)

1°

The backing off angle of 1° appears to have been arrived at somewhat arbitrarily but has proved successful for this ratio of sizes.

Noise reduction 8–10dB(A)
Cost: negligible

Noise reduction at ring spinning frames

Background

Ring spinning frames are widely used in the manufacture of yarns and threads and results reported in the First Report of the Joint Standing Committee on Health and Welfare in the Cotton and Allied Fibres Industry showed that noise levels at these machines varied from 91 to 102dB(A) depending on the product; the condition and age of the machine; the number of machines in operation and the acoustic conditions of the installation. These machines may have up to 150 spindles running at up to 15 000 rev/min. The major sources of noise and modifications which have been effective in reducing noise on new machines are shown below.

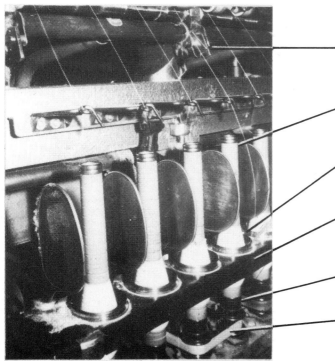

Showing major sources of noise and associated modifications

Noise Sources	Modifications
Vacuum and collection	Fit silencers
Bobbin and spindle (125Hz-250Hz)	Use well fitting and balanced bobbins
Ring and Traveller (4kHz-10kHz)	Fit isolating, elastomeric ring holders (Figure 2)
Ring carrier	Isolate ring carrier from machine frame
Tape spindle drive	Use narrow tape drives
Spindle bearings (500Hz-4kHz)	Isolate bearings from their holders and the the spindle rail

Tests on multiple machine installations of machines incorporating the above modifications gave noise levels in the range 87–90dB(A) for machines operating at spindle speeds of 13 000 rev/min.

Limitations

The above modifications have already been incorporated in the design of new machines but no retrofitting has yet been carried out in this country. Sources in the USA claim that spindle bearing mounts and traveller ring isolators can be retrofitted.

Showing rings mounted in elastomeric isolators

Claimed noise reduction 4–12dB(A)
5dB(A) (isolated bearings and ring holders)
Estimated cost (1981) £10–20 per spindle (isolated bearings and ring holders)

Press maintenance and tool design

A study of press working noise by the Institute of Sound & Vibration Research has resulted in various recommendations to reduce the generation of hazardous noise.

As part of this study a detailed investigation* was made of the effect mechanical condition of the press can have. Measurements were made of the process noise as a simple 20 ton 'C' frame press pierced 20mm diameter holes in aluminium plates 3 and 6mm thick with a flat punch and die.

The press was then refurbished with new bearing shells, properly adjusted slide bearings and oil flood lubrication. Identical piercing operations were repeated for direct comparison which showed reductions at the operator position between 7 and 16dB(A) depending on thickness of plates and the details of the tool conditions.

Another aspect of the overall study was concerned with punch:die clearance and its effect on noise generation. The rebuilt press showed much greater benefit from the 'quieter' (low clearance) condition in the tool and would be expected to respond similarly to other low noise tool design features e.g. Shear punch.

Other advantages: Restoration of capacity. It was noted that after the maintenance work the press was able to exert its rated tonnage, whereas previously it only reached 60% of this figure.

Accessibility unaffected.

C-frame press

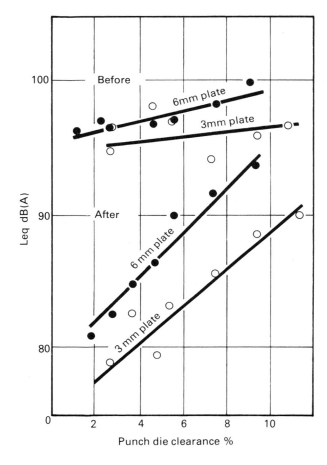

Noise piercing aluminium plate before and after press rebuild

Claimed noise reduction: 7 to 16dB(A)
Estimated cost (1981) £1,000

* Further details: A G Herbert, Wolfson Unit for Noise and Vibration Control, Institute of Sound and Vibration Research, University of Southampton

Noise reduction treatment of a fork lift truck

Background

Investigations* have defined the predominant noise sources contributing to excessive levels in several fork lift trucks. Such investigations have provided retrospective solutions to the problem in the form of palliative noise reduction packages. The details of each noise reducing solution will be peculiar to a particular type of lift truck, however the following is given as a guide to the treatment which is considered to be appropriate and to illustrate what reductions in noise levels at the lift truck driver's ear position may be achieved.

For information, details are given for both a cabbed lift truck and for a lift truck with protective case only.

Noise control treatment

Noise reduction at driver's ear (dBA) Areas of treatment	Absorptive head lining in cab; noise reducing floor mat; improved acoustic sealing of: bonnet, lower edges of cab doors, cab front area forward of hand control console; damping/absorptive lining on underside of bonnet	Noise insulating material applied to bonnet area	Modified cooling air passages and improved engine door seals	Reduction of engine r.p.m. where possible
Uncabbed truck	2	2	5	1/100 r.p.m.
Cabbed truck	3	3	16 (ducted out of cab)	

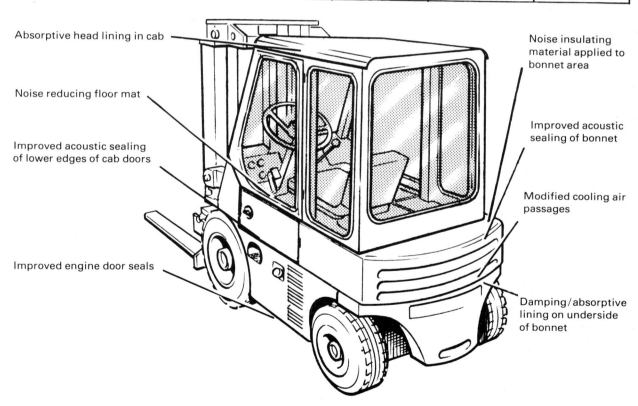

Absorptive head lining in cab

Noise reducing floor mat

Improved acoustic sealing of lower edges of cab doors

Improved engine door seals

Noise insulating material applied to bonnet area

Improved acoustic sealing of bonnet

Modified cooling air passages

Damping/absorptive lining on underside of bonnet

Claimed noise reduction 22dB(A) Cabbed truck
9dB(A) Uncabbed truck
Approximate cost (1981) £350–£450

* S.R.L. Ltd, Sudbury, Suffolk

Air cooled diesel trucks

Background

Many types of construction plant emit excessive levels of noise. The need for quieter plant is becoming an urgent matter particularly since EEC directives are in preparation which will limit noise emission from construction plant both stationary and in motion.

Noise reduction treatment

Investigations* on a 13·4kW dumper† with an air cooled diesel engine have shown that a worthwhile noise reduction can be achieved by applying recognised methods for noise suppression. Considerable reductions were achieved by fitting more efficient exhaust and inlet silencers and by enclosing the engine. (Particular attention had to be paid to the size of ducts and apertures to obtain sufficient engine cooling). Such noise reduction treatments could be applied to other plant powered by air cooled diesel engines, notably rough terrain fork lift trucks and concrete mixers.

Site trials

After noise reduction treatment the dumper was subjected to 3 months site trials during which no problems were experienced and there was no increase in noise at the end of the trial period.

Dumper in standard form

The modified dumper with the engine canopy hinged showing noise absorbent lining

Noise reduction

Noise treatment	Noise reduction at driver's ear (approx)
Replacing 'pepperpot' exhaust and standard air filter with acoustic mufflers and filter	2–4dB(A)
Extend and reroute tailpipe	1dB(A)
Acoustic enclosure of engine	4dB(A)

Claimed noise reduction 8–10dB(A)
Cost (1981) 7% over conventional truck

* All stages of the project described in BRE papers: CP 32/76 and CP 40/77 obtainable from Distribution Unit, Building Research Establishment, Garston, Watford
† Acknowledgements to Babcock Construction Equipment Ltd, Gloucester

Noise reduction at jolt squeeze machines

A company specialising in non-ferrous castings were faced with two problems during machine moulding using jolt-squeeze machines. The first problem was one of mismatch of the castings caused by deformation of the mould due to long vibration sequences needed to free the mould. Also because of the hammer action of pneumatic piston on the mould base the pattern of vibration was in one direction. Enquiries by the company revealed that an alternative vibrator was available this being a rotary vibrator*. The action of this vibrator was to force a steel roller round a circular track by com-

pressed air, the frequency of vibration could be controlled by controlling the air pressure and fixing to the mould plate was similar to that of the conventional vibrator. Trials revealed that the vibration pattern was multidirectional which resulted in better mould-from-pattern release and the required vibration period was shorter. In addition the amplitude of vibration required was lower. Secondly the noise level during vibration was considerably lower.

The photographs below show both types of vibrator.

Conventional pneumatic vibrator
Maximum sound level – 110dB(A)

Rotary vibrator
Maximum sound level – 95dB(A)

Showing the rotary vibrator attached to mould place
Photographs by kind permission of R H Roseblade & Son, London NW10

Noise reduction 15dB(A)
Cost of rotary vibrator (1980) £200 (pair)
Cost of pneumatic vibrator (1980) £170 (pair)

* Rotary vibrator supplied by Vibrotechnique Ltd, Brighton, Sussex

Noise reduction at bar reeling and straightening machine

A company specialising in production of high speed and tool steels were faced with the problem of very high noise levels produced by a Robertson bar reeling machine. Levels of noise between 97 and 102dB(A) were measured at both the operator 'in-feed' and 'offtake' positions and many other people in adjacent work areas were also exposed to excessive noise levels from the reeling operation.

The following sources of noise were identified at the reeling operation:
(a) The input bar feed stock tube
(b) The various guide tubes on the reeler
(c) Handling of metal rods before and after reeling
(d) Pneumatic exhaust noise on various controls and actuators.

The company have very successfully solved the noise problem by the following means:
(a) Construction of a novel bar feed stock tube
(b) Design and provision of an acoustic enclosure around the bar reeler
(c) Lining of metal racks and output rod container with shock absorbing rubber
(d) Fitting of pneumatic exhaust silencers.

The enclosure illustrated below is in three parts, viz:
(a) A permanently located skirt
(b) Two top parts, one of which can be speedily removed for roll changing.

The complete enclosure is shown in Fig 3 and incorporates visual access at the inlet side and an access door for altering roll clearances. The levels of noise at the in-feed and offtake operators' positions are now 82 and 84dB(A).

Permanently located skirt

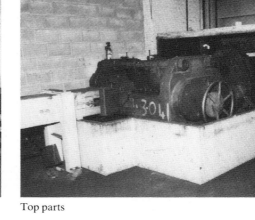

Top parts

Complete enclosure

Photographs by permission of F M Parkin (Sheffield) Ltd, Sheffield

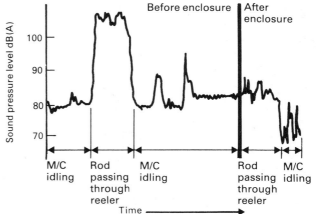

Noise levels before and after enclosure

Noise reduction 20dB(A)
Approximate cost (1981) £10,000 (including design and development costs)

'Arcstract' system reduces fume and noise from air-carbon arc gouging

The high metal removal rate achieved by air-carbon arc gouging compared with more traditional fettling methods has made it a particularly attractive process for fettling steel castings. In basic terms it consists of striking an electric-arc between a carbon electrode and the steel casting and removal of the molten material using a powerful blast of compressed air. However, accompanying the process are unwanted environmental problems of fume, noise, heat, ultra-violet radiation and metal splatter. Investigation and development engineering by a foundry research association* has provided a fully integrated solution to these problems with a completely new type of air-carbon arc booth.

The booth is available in two styles as shown above.

The fixed version requires small castings to be lifted into the booth through a small opening at the front. A turntable eases handling inside the booth.

The sliding version is for gouging work on larger castings, up to a cross sectional area of 4ft × 2ft and long sections of steel. A motor driven unit enables the booth to be moved along the entire length of the work stage by stage.

Normal arcing takes place inside the extraction booth which protects the operator from excessive heat, molten metal splatter, fume and noise. The internal surfaces of the booth are acoustically lined to prevent noise build-up. Clear visibility of the process is provided by inclined double glazed observation windows which also give protection from ultra-violet radiation. A tinted screen of variable position is fitted to the panel to permit observation of the arc but leaves both hands free.

Fixed booth
Photograph by permission of Parker Foundry Ltd, Derby

Sliding booth
Photograph by permission of Ryder Bros Ltd, Bolton

Claimed noise reduction 10dB(A)
Approximate cost (1982) £4,000 (fixed booth with 3–4000cfm ext)

* Steel Castings and Trade Research Association, Sheffield
SCRATA video available title 'Arcstract Systems in Steel Foundries'.

Wire bunching enclosure combining acoustic and safety requirements

Wire bunching machines are capable of emitting considerable levels of noise generated by this notoriously noisy process. Machines manufactured by Hungarian Cable Making are marketed in Western Europe by a German company. These machines were fitted with safety guards as standard, however high noise levels were emitted. A British company* were commissioned to design and manufacture alternative guarding enclosures which would combine both noise control and adequate safety guarding. These enclosures are now fitted as standard to both the models of machine supplied, DSO 40 and DSO 63 and can be retrofitted to existing machines.

The enclosure incorporates a sliding access door with integral shatterproof window, forced draught ventilation and safety interlocks. It can be lifted clear of the machine in one piece for conducting major maintenance or, if no overhead lifting facilities are available, can be easily taken apart since it is constructed from several independent panels.

In the foreground above is a machine fitted with the new enclosure. In the background can be seen a machine with the original safety guards.
Photograph by permission of Berma Maschinehandelsgesellschaft m.b.H.

Claimed noise reduction 27–30dB(A)
Cost (1981) £1725

* Noise Control Centre, Melton Mowbray, Leics.

An integral folder enclosure

Measurements have shown that the greatest contributor to printing room noise levels is the folder with, at speeds above 30 000 copies per hour, most noise being produced by the folding blade contacting the paper being folded and a subsequent folded edge slap at the second fold rollers. Noise from the print units and the Kite former, is also significant and restricts the extent to which action to reduce folder noise can reduce the noise exposure of machine operators.

One way of reducing folder noise and at the same time providing guarding is to enclose the folder as shown below. The enclosure* is fitted to a Goss Urbanite, single width printing press rated at 40 000 off, 32-page copies per hour and is constructed from sound absorbent lined steel panels.

Access doors are fitted on both sides of the folder and are interlocked by means of a Castell Key system which limits the drive to a crawl speed when the door key is removed from the drive electrical interlock panel. The access door on the operator's side is of transparent plastic.

Quick release air extraction hoods, coupled to underfloor ducts are mounted inside the enclosure. A short acoustic tunnel is fitted to limit noise breakout from the take off conveyor aperture.

NB. The noise reduction values given below are the results of measurements taken at fixed positions on machines with enclosed and machines with unenclosed folders. Because of the presence of noise from other sources the values do not truly reflect the maximum extent to which the noise generated by the folder has been reduced by enclosure.

Photograph by permission of North Western Newspaper Co Ltd, Blackburn

Noise reduction (at folder man's position) 5 to 6dB(A)

* Supplied with the machine by the manufacturer MGD Graphic Systems Ltd, Preston, Lancs.

Close shielding on a power press

Background

30-ton power presses were used for continuously fed simple blanking. Major noise sources were, therefore:
(a) Pneumatic, from air operated blow off.
(b) Impulse from the blanking operation.
(c) Impact, from components ejecting into bins.

The machines run between 1 to 3 strokes per second with feed material replacement between $\frac{1}{2}$ and 1 hour.

Modifications

1. Main acoustic enclosure, interlocked to form press guard.
2. Low voltage light inside enclosure.
3. Rear acoustic cover completes enclosure.
4. Attenuated slot for component ejection.
5. Hinged plastic flap on collection bin.
6. Aerodynamic nozzle for blow off operate intermittently as required, reducing air pressure from 80 to 20 psi.
7. Silencer fitted to pneumatic exhaust.
8. Vibration isolators under press.

Graphical levels show reductions from 111d(B(A) to 90dB(A) with mid-stage (without reduced air pressure) of 97dB(A). These cannot be considered average operator levels since operators move about. Closer figures would be taken at 1m. The estimates at this point are the ones quoted i.e. 105dB(A) to 84dB(A).

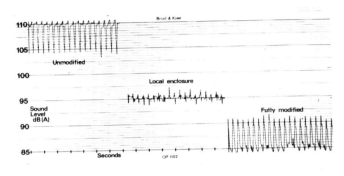

An unmodified press 105dB(A)
Photograph by permission of GKN Floform Ltd, Welshpool

A modified press 84dB(A)

Noise Reduction 21dB(A)

Enclosure of cold heading machines

Cold heading is a notoriously noisy process due not only to the impact of cold forging, but also to the structure borne excitation of the machines structure. To the noise resulting from these sources, must be added that due to mechanical drives and other machine services. These machines are often used in large numbers which results in excessively noisy work places, the levels being typically 105–110dB(A).

Several means of noise reduction are being attempted, and some benefits can be expected with the use of compressed air and slider preloading, but at the moment the only way to achieve acceptable attenuation is by complete enclosure.

The teleclosure systems, illustrated below, have been designed to give high noise reduction whilst providing total access to the machine. They can be completely removed or quickly replaced with the minimum of effort or interference with machine services.

Additional advantages are that, suitably interlocked, they provide secure fencing and increase the efficiency of oil mist extraction.

Claimed reduction 15–25dB(A), dependant upon machine service requirements
Approximate cost (1980) £1500–£1600

Teleclosures by The Noise Control Centre, Melton Mowbray, Leics

29

Acoustic hopper for plastic granulator

Grinding plastics produces high levels of noise not only due to the action of grinding but also due to the vibration excitation of the large metal surfaces, such as those often forming the throat and hopper, due to impact of the plastic.

An internationally known company purchased five mobile granulator units, for use by operators on the plastic moulding processes to grind the plastic offcuts, 'sprew'. The granulators were fitted with hoppers of 10 gauge mild steel as standard. Feeding of the sprew into the mouth of the hoppers proved somewhat difficult, also noise levels of up to 95dB(A) were measured at the operator position.

It was decided to modify the hopper design to ease loading and also at the same time to incorporate acoustic treatment to reduce operator noise exposure.

The modified hopper, shown in the photograph above attached to the granulator, was constructed of a double skin 16 gauge steel with a 25mm cavity filled with acoustic tile. A cover was incorporated over the loading aperture which was lifted when loading but prevented noise escape from the aperture when in place. Noise levels at the operator position when the granulator was fitted with this new design of hopper were reduced to 83dB(A).

Photograph by permission of Biro Bic Ltd, London NW10

Noise reduction 12dB(A)
Estimated cost (1981) £20

Book binding machine

Oxford University Press use a bookbinding machine which can generate hazardous noise levels when its book trimming facility is used: the backs of papers are sliced across by a rapidly rotating circular saw to give a neat finish prior to binding.

A simple and efficient solution to the noise problem was achieved by lining the inside of the existing hinged cage guard over the trimming zone with 2 in thick expanded polyurethane foam.

As shown below, one metre in front of the cutting zone, sample Leq during trimming was reduced from 95dB(A) down to 85dB(A).

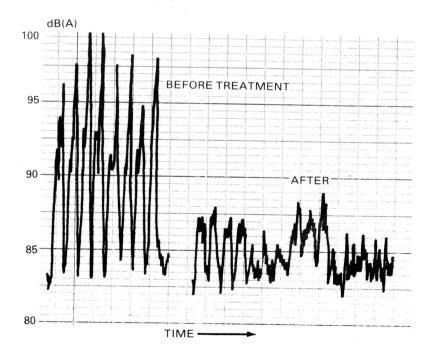

Noise reduction 10dB(A)
Estimated cost £5

* By kind permission Oxford University Press, Printing Division, Oxford

Acoustic guard for paper bag making machine

Very high noise levels, up to 110dB(A) continuous, can be generated by paper bag making machines. Noise levels vary, mainly due to the speed at which the machine is run, up to 1000 bags per minute, and to a lesser extent depending upon the size and quality of the paper used.

The principal noise source on this machine is the impact of a rotating snatch block as it tears the paper sheath to a bag length against a synchronised roller fitted with a serrated knife.

To minimise the noise from this source and also improve the guarding of the dangerous parts of this machine, a well established firm of paper bag manufac-

turers designed and fitted the acoustic guard illustrated below. The guard was manufactured from 5mm thick acrylic sheeting (Micor) carefully contoured to fit the machine casting. Easy access to the machine parts is provided by hinging this cover as shown with an electrical interlock arrangement. The outfeed area of the machine is treated in a similar manner.

Tests undertaken on comparable machines running at 900 bags per minute show that the acoustic guard reduced the noise level from 112dB(A) above the snatch block to 104dB(A) and from 105dB(A) to 99dB(A) at the outfeed area (usual operator position).

General view of the machine fitted with acoustic guard.

Acoustic guard in open position

Photographs by permission of Dolans Corrugated Containers Ltd, Bristol

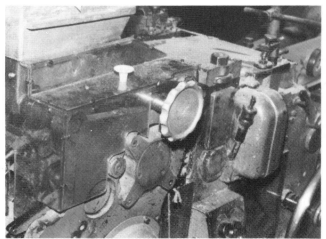

Acoustic guard at outfeed.

Noise reduction – 6 to 8dB(A).
Estimated cost (1981) £100 per machine

C-Frame press

The consultative document *Protection of Hearing at Work** and its companion background document. *Some Aspects of Noise and Hearing Loss†* both included a drawing of an acoustically treated 'C' frame press. They were included to show typical noise sources and how reduction treatment can combine safety guarding and acoustic close shielding.

Only the front illustration of the press was used but art work had been prepared for a rear view also. Both are, therefore, reproduced here for the convenience of inspectors.

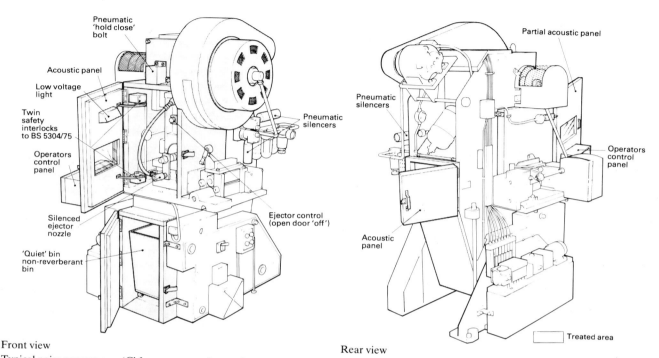

Front view

Rear view

Typical noise sources on a 'C' frame press and some forms of treatment including the combination of safety guard and acoustic close shielding.

Noise reduction (Actual installation 21 dB(A)
Estimated cost (in-house) £200–£400

* HMSO £3·00 net (ISBN 0 11 883431 2)
† HMSO £3·50 net (ISBN 0 11 883432 0)

Acoustic hood for 'Monotype' composition caster also acts as a safety guard

Composition casters are frequently the source of high noise levels in many printing plants. The Monotype Corporation Ltd*, aware of duties under Section 6 of the Health and Safety at Work etc Act (1974) was concerned at the noise levels generated by their machines and investigated methods to reduce the noise. An acoustic hood was developed which may be easily retrofitted to their existing machines to reduce the noise levels considerably. The hood is constructed of 18g mild steel, lined with sound absorbing material which is protected from oil, dirt and damage.

A hinged window of 6mm perspex allows access and adequate viewing of the working zone and other hinged access flaps are placed at strategic positions defined, after exhaustive trials, to provide ready access for running adjustments and routine maintenance. The complete hood is supported on 'slide-off' hinges to provide quick access and removal if necessary for major maintenance.

Photograph by permission of London College of Printing London EC1

Noise Reduction 5–6dB(A)
Cost (1982) £1020

* Monotype Corporation Ltd, Redhill, Surrey

Saw blade dampers

Noise from circular sawing machines will depend not only on the characteristics of the material being cut but also on the rate it is being cut and the mechanical characteristics of the cutting saw. Noise sources may include: impact noise of the saw against the workpiece, vibration of the workpiece, aerodynamic noise from the saw blade, and also vibration of the saw blade, together with many other noise sources associated with the machine.

Aerodynamic saw blade noise (hiss) will be considerably reduced by running at the correct blade peripheral speed. This is usually the most predominant source of saw blade noise. However, occasionally, particularly with large diameter saw blades (over 400mm diameter), high levels of blade vibration occur which result in tonal type noise (ringing). Plate saw blades are more susceptible to vibration than tungsten carbide tipped blades. Such vibration may be reduced by increasing the stiffness and damping of the saw blade.

A damper disc has been designed which, when correctly fitted to one surface of the saw, should reduce blade vibration generated noise levels considerably. The disc is a 10 thou thick tin plated steel disc coated with a 2 thou viscoelastic adhesive. As the blade and steel disc distort under vibration, the viscoelastic layer sandwiched between the two, is deformed. This converts the

mechanical energy of vibration into negligible heat which is dissipated within the damping layer.

N.B.: The damper disc may have to be removed for blade retensioning or when the blade is heated dependent upon procedures used. In no case should a damaged disc remain on the blade.

Machine: Wadkin Circular Wood Saw **Blade:** 30 in diameter, 1 in gullet depth, 50 teeth, motor speed 1445 rpm **Damper:** 22½ in (largest available) **Microphone Position:** 1 – operator position at feed end 2 – 2 ft to side along axis of blade

Noise levels

Material being cut	Condition	Sound level dB(A)				Effective noise reduction dB(A)	
		Position 1		Position 2		Position 1	Position 2
		Undamped blade	Damped blade	Undamped blade	Damped blade		
— 3″ thick Jelutong' Softwood	Idling Cutting	93 97	93 94	93 102	90 93	0 3	3 9
2″ thick Afrormosia Hardwood	Cutting	101	98	106	98	3	8

The above noise levels were obtained on a saw which had very intermittent use in a pattern shop*. The practical difficulties which may be encountered when using saws fitted with damping discs for long periods of time such as on production work are not known. It is recommended that such treatment should be first applied on a trial basis to assess any difficulties which may occur particularly during blade retensioning.

Noise reduction 3dB(A) at operator feed position
8–9dB(A) along axis of saw blade
Cost (1981) 25%–40% increase on cost of blade

* By kind permission of Rolls Royce, Bristol.

Damped component chute

A centreless grinder was found to be producing excessive noise. Analysis of the noise showed that the sound of the ejected product falling in the exit chute was significantly responsible for the high noise level. Replacement of the metal chute by one of Ultra High Molecular Weight Polyethylene reduced noise to an accepted level.

As shown below sample Leq was reduced from 90–92dB(A) using the metal chute to 81–82dB(A) with the plastic chute.

'Before'

Metal chute

'After'

Ultra high molecular weight plastic chute

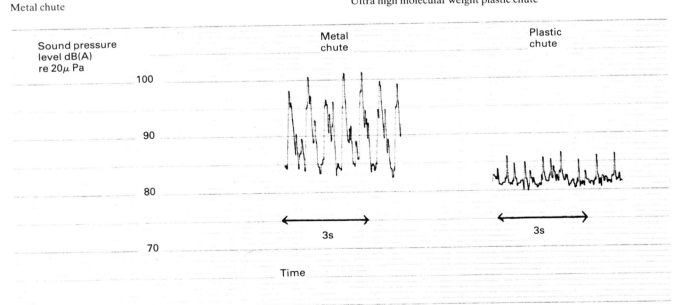

Additional advantages

Safety	– wear does not produce sharp edges
Performance	– reduced component damage
Lifetime	– greater resistance to wear
Noise reduction	– 5–10dB(A)
Approximate Cost	– less than half price of metal chute

Suppliers of U.H.M.W.P. *Performance Plastics*, Bacup, Lancs. *I.A.C. Plastics*, Burnley, Lancs.

Cushioning of metal-metal impact by rubber lining

A potential noise hazard was identified in a chicken hatchery's tray washing room. It was found that the hazard was dominated by metal-to-metal impact as metal hatching trays were banged against a hopper to remove waste and as they were loaded onto a metal base plate at the input to the washing unit.

The firm was advised to reduce this impact noise by covering the metal surfaces at the hopper and washing unit which were struck by the trays with a wear resistant cushioning material. The firm called in a local agricultural engineer* who bolted 6mm thick cord-reinforced rubber mat (as used for flexible doors) onto the required surfaces, shown below. This provision reduced the operator's exposure to noise levels from 93dB(A) to 87dB(A) Leq (8 hour).

Left, washing unit input, right waste hopper
Photographs by permission J. P. Wood & Sons (Hatchery), Craven Arms

Rubber mat applied to
lip of washing unit input

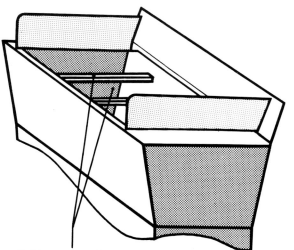

Rubber mat applied to
inner hopper surfaces
and to top of horiz-
ontal bumper bars.

Noise reduction 6dB(A)
Cost (1981) less than £100

* Noise treatment by R. Freeman, Agricultural Engineer, Craven Arms

Damping of vibrating hopper

Prior to packaging or 'jarring' of hard boiled sweets they are conveyor fed to drop into the hoppers shown below. The hoppers are vibrated to assist the natural gravity feed of sweets to the discharge chutes and the containers. Dropping sweets and the vibration of the hoppers can create high noise levels.

To reduce the noise new hoppers made of sound deadened steel were introduced and also the impact surfaces were lined with thick plastic sheeting, see photograph below. Noise measurements prior to installation are put at 97dB(A) and after installation 88dB(A). It is believed that the plastic sheeting alone reduced impact noise by 3dB(A).

Claimed noise reduction 9dB(A)
Cost (1981) £300 per hopper

Information by courtesy of Cadbury Ltd, Keynsham, Bristol

Sound deadened table

In many spirits and fruit bottling processes new bottles arrive at the bottling hall in the same containers used to package the filled bottles. The new bottles are emptied from the container by inverting it by hand and dumping the bottles onto a table adjacent to an unscrambling device which transfers them to the main conveyor line. This dumping creates high impulsive noise levels.

Noise control

As part of a bottling hall noise control programme a company treated their existing wood dumping tables with acoustic material to deaden the high impulsive noise. The tables were covered with a flexible plastic material which was backed with self adhesive polyurethane foam with a damped coating.

Using the treated table the peak levels of impulsive noise were reduced by up to 12dB(A) which reduced the noise exposure for the operator at the table from sample Leq 94dB(A) to 90dB(A).

Sound deadened table
Photograph by permission John Haig & Co Ltd, Glasgow.

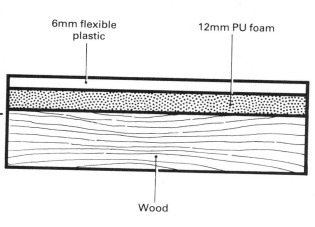

Sectional structure of acoustic treatment

Noise reduction 4dB(A)
Approximate material cost £10 per square metre

Noise reduction at minimal cost

Vibration is often applied to moulds containing concrete to aid setting and extraction of air. A company manufacturing reinforced concrete products successfully reduced noise from the vibrating table on which the moulds of concrete cubes were placed, with a minimum of expenditure. Noise was generated by the steel moulds bouncing on the steel surface of the table, there was also noise from the vibrator itself. A considerable reduction in noise was achieved by fixing a sheet of 3/8″ rubber to the steel surface of the vibrator platform. Also similar rubber was used to make a skirt for the vibrator so preventing noise generated beneath the platform by the vibrator from reaching the operator. The rubber used was scrap conveyor belting rubber.

Table before treatment – 106dB(A)

Photograph by permission of Anglian Building Products Ltd, Norwich

Table after treatment – 93dB(A)

Noise reduction 13dB(A)
Cost NIL (scrap material used)

Noise reduced when hammering metal section

A common noise problem in metal fabrication work-shops is the high impact noise levels generated when hammering metal sections.

A company engaged in the production of a wide variety of fabricated metal products faced with such a problem when hammering metal sections to straighten them prior to fabrication, used some initiative and achieved a reduction in these high impact noise levels.

Steel sections were straightened prior to welding on blocks constructed from steel channel forming the base, an RSJ forming the stand, and square section steel forming the blocks.

High impact noise levels were reduced by:
(i) filling the steel channel base section with concrete
(ii) welding 16g steel plate across the flanges of the RSJ and packing the voids with glass fibre wool.

Using the treated straightening blocks the peak levels of impulsive noise were reduced which reduced the noise exposure of the person hammering from sample Leq 102dB(A) to 96dB(A).

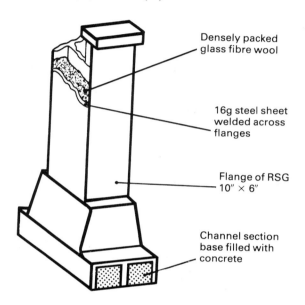

Photograph by permission of Singer and James Ilford.

Densely packed glass fibre wool

16g steel sheet welded across flanges

Flange of RSG 10" × 6"

Channel section base filled with concrete

Noise reduction 6dB(A)
Cost (1981) £5–£10

Flywheel damping

Certain components within mechanisms can be major radiators of noise, large flywheels of power presses being typical examples. Vibrations of the flywheels of power presses may be excited both by operation of the clutch and by the impact of the tooling itself. The noise is characteristically pure tone, the frequency depending on the modes of vibration excited in the flywheel.

Analysis of vibration patterns of the flywheel of a 75-ton press, followed by computer modelling of the modes of vibration, allowed tuned dynamic absorbers to be designed. These consisted of six sets of two steel plates bolted together through the flywheel with a layer of rubber bonded cork between each plate and the flywheel.

The intrusive pure tones in the noise spectrum were reduced by 11dB(A) at 1250 Hz resulting in an overall reduction of 4dB(A).

The diagram shows details of the tuned absorbers used in the treatment. The precise measurements given have been calculated for this individual machine and have only been given as an indication. **Similar measurements will not be appropriate to other machines** and must be re-calculated.

This technique can be applied in many cases where flywheel resonances dominate noise radiation, using the same modelling techniques for flywheels which radiate at other frequencies to calculate the necessary dimensions of the appropriate tuned absorber.

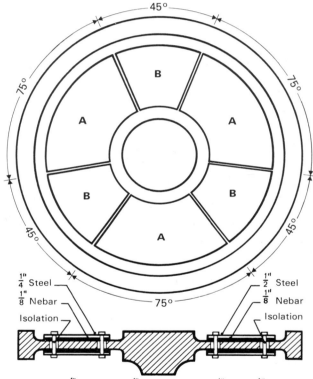

Sandwich A : $\frac{1}{2}$" Steel Plate/ $\frac{1}{8}$" Nebar/Flywheel/ $\frac{1}{8}$" Nevar/ $\frac{1}{2}$" Steel Plate
Sandwich B : $\frac{1}{4}$" Steel Plate/ $\frac{1}{8}$" Nebar/Flywheel/ $\frac{1}{8}$" Nebar/ $\frac{1}{4}$" Steel Plate

To avoid vibration transmission all surfaces should be fixed with an adhesive. The bolts are an added safety feature only and, therefore, should not be in contact with the flywheel.

Claimed noise reduction 4dB(A) (11dB(A) at resonant frequency)
Estimated cost (1981) £400 (included consultancy and installation of tuned absorbers as above)

* Lucas Industries Noise Centre, London W.3

Noise reduction of rotary tumblers

A firm who manufacture chains has achieved considerable noise reduction at their four rotary tumblers used as part of the chain cleaning process. High noise levels, sample Leq 108dB(A), were generated as the chains tumbled inside the original resounding tumbler boxes which were constructed of 4·5mm steel plate.

Heavier tumbler boxes have been constructed of 12mm thick welded steel sheets. Vibration damping to further reduce noise was achieved by welding smaller 6mm thick plates onto the four large faces of the box as shown below. (NB. The centre spot weld is important to achieve sufficient damping) The new tumbler boxes achieved a noise reduction of at least 7dB(A).

Existing partial enclosures segregating the adjacent tumblers were converted to full sound reducing enclosures (1·5m × 1·3m × 2·4m high). The enclosures used an existing rear brick wall, the sides and roof were 110mm thick comprising 19mm chipboard outer surface with 12mm fibre board inside sandwiching compressed fibre-glass blanket, all mounted on a framework of 75mm × 50mm wood. The inner surfaces were lined with 50mm thick mineral wool to reduce reverberant sound. Access was via sliding doors 80mm thick constructed of two 25mm plywood sheets sandwiching fibre glass blanket. These were suspended from an overhead rail and sealed with the floor with a brush-type draught proof strip. Doors sealed tight when closed by the method sketched below.

New tumbler box with increased vibration damping. Noise reduction 7dB(A)

Photographs by permission of William Hackett Ltd, Cradley.

Enclosure. Noise reduction 20dB(A)

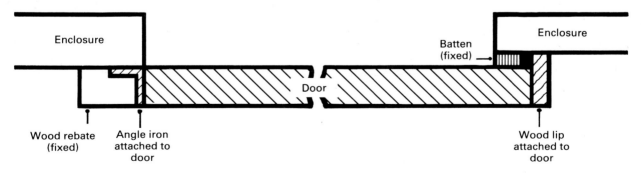

Noise reduction 27dB(A) (7dB(A) due to vibration damping)
Cost (1981) £2000 (installed for four machines as above)

Plastic gears reduce noise of beck sheeter

The machine is used for cutting paper into sheets (maximum size 4 × 4 ft). It was purchased second-hand in 1975/76 and can be seen in the photograph below. The machine is powered by a 7·5 hp electric motor and the maximum rate of production is 100 metres per minute.

During a noise survey by the company it was noticed that the noise level at the operator's position was 93–94dB(A). It was established that a principal noise source was the steel on steel gearing and a possible solution was to replace intermediate gears from steel to plastic. The company engaged a local plastics firm† to manufacture the new gears of black nylon impregnated with molybdenum (MOS 2). After the gear change the noise level was reduced to approximately 85dB(A).

(a) Plastic gear as fitted for noise reduction (b) General view of the machine
Photographs by permission of Aarque Systems Ltd, Colnbrook Slough

Approximately noise reduction 7–8dB(A)
Cost (1981) £50

† Iver Plastics Ltd, Langley, Berkshire

Quiet nozzles reduce air ejection noise

Noise due to air ejection of components often represents a significant contribution to the total noise dose of a machine operator. Noise from ejection air jets can be reduced by any of the following measures:

a) Make a continuous jet interrupted to be the only available when required
b) Fit acoustic quiet nozzles
c) Reduce the duration of the jet
d) Reduce the air pressure

Combinations of these measures can be found which achieve not only effective ejection and an overall reduction in noise but also a considerable saving in compressed air.

A machine manufacturing company successfully reduced the noise for a 50 ton automatic Hydrena fine blanking press stroking at 45 strokes per minute from 99dB(A) to 93dB(A) sample Leq by a combination of the above measures a) and b). They fitted two Schrader* Bellows 401 quiet nozzles to the two air ejection nozzles and found that not only was there a reduction in noise level but also, due to the improved aerodynamic performance of the nozzles, the jet period needed to remove the piece part of 13g steel effectively was half that for the previous plain tube jet thus giving a saving in compressed air.

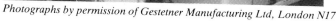

Photographs by permission of Gestetner Manufacturing Ltd, London N17

Noise reduction 6dB(A)
Cost (1981) less than £20

* Schrader Bellows Ltd, Bridgtown, Staffs.

Low frequency boiler noise

Background

Low frequency noise generated by four industrial boilers housed in a hospital boiler house produced complaints from the owner of an adjacent residential property some 50m away. The boilers were forced draught fan boilers with rotary cup burners, and each had already been fitted with high performance absorptive silencers on the forced draught air intake. The nature of the noise problem was a low frequency tonal peak which was apparent throughout the adjacent property when any one of the boilers was on a low setting. The level of the peak varied considerably, however the frequency varied only between 30 and 33Hz and was audible for long periods, including throughout the nighttime period, whilst any one of the boilers was on a low setting. Within the boiler house the reverberant level of this peak was 85dB(Lin); outside the complaintant's house the level fell to 79dB(Lin) and within the house levels of 67 and 62dB(Lin) were found in the lounge and bedroom respectively.

Noise control

Investigations by a specialist consultant firm indicated that the source of the noise from the boilers at low fire settings was air turbulence. The forced draught air intake was controlled by dampers with the fan operating at full speed, but caused high air turbulence at the low fire settings and radiation of the low frequency peak.

After various tests it was concluded that a reactive Helmholtz silencer would be needed to bring about the significant low frequency noise reduction necessary for complaint to cease. Such a silencer, allowing volume changes for fine tuning once in position, was designed and installed. It produced a reduction at the frequency in question of 30dB in both the boiler house and within the adjacent dwelling.

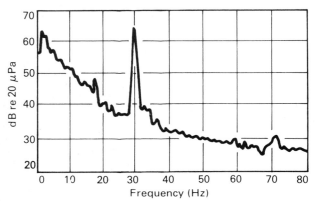

Typical noise spectrum of offending noise in house

Silencer installation on a boiler

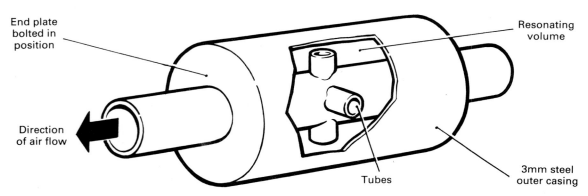

Cutaway view through resonating silencer

Noise reduction 30dB (at approximately 30 Hz)
Cost (1981) £5000 (included consultancy and installation of four silencers as above).

* Thanks are due to Dr F. Chatterton (Acoustics Group) Environmental Planning and Design Ltd, Basingstoke, Hants.

Compressor silencing

High levels of low frequency noise generated by compressors caused vibrations of windows and thin panels and also considerable disturbance to workers in an office about 10m away.

Noise and vibration measurements indicated that the pulsations from the compressor intakes at the firing frequency of 25 Hz were causing the noise.

The solution* to the problem was to fit 'off the shelf' reactive type silencers downstream of the air-intake filters. This reduced the low frequency noise by 10dB and made a considerable reduction in subjective annoyance. The silencers† were designed and manufactured for this specific problem.

Noise levels were measured in octave bands with the overall A-weighting at the doorway of the compressor house with compressor running on and off load.

An interesting point that emerges from this case is that, being a low-frequency noise problem, there was no change in A-weighted overall noise level before and after the silencers were fitted. This shows that although dB(A) is nowadays used for assessing whether a noise might cause risk of hearing damage, dB(A) may not be a reliable guide to the annoyance caused and the character of the sound will then need to be taken into consideration as well as its level.

Photograph by permission of UK Provident Life Assurance, London

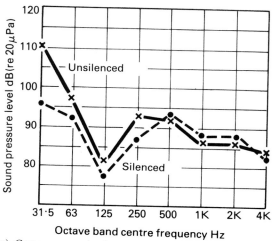

a) Compressor on load
Silenced: 94dB(A), 100dB(lin)   Unsilenced: 93dB(A), 110dB(lin)

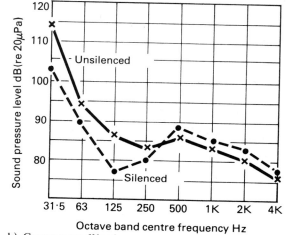

b) Compressor off load
Silenced: 90dB(A), 103·5dB(lin)   Unsilenced: 90dB(A), 114dB(lin)

Claimed noise reduction 10dB(lin)
Approximate cost (1981) £2,000 (Five units treated)

* Wolfson Unit for Noise and Vibration Control, University of Southampton, Hants.
† Moniton Technic Ltd, Basingstoke, Hants.

Reciprocating compressor noise

Reduction of low frequency pulsations from the reciprocating compressor poses a particular problem. Hearing protection is seldom effective at such low frequencies (below 50Hz) and the noise must therefore be reduced at source. Absorption type silencer units which readily reduce high frequency noise of fans and axial flow compressors are ineffective at such low frequencies, however a considerable reduction in such noise has been achieved with acoustic inertia damper type silencers*. The noise reduction characteristics of one such series of dampers is achieved entirely by correct design of integral resonators and expansion chambers without recourse to absorption materials. This minimises the danger of foreign matter being drawn into the compressor. The dampers are aerodynamically designed to effectively cope with the two dimentional air flow encountered at inlet to reciprocating compressors, and eliminates the associated high pressure drop. The dampers are installed directly onto a compressor inlet or at the end of a pipe run. Noise reductions of 10 to 20dB have been obtained at the lowest discrete frequencies on an octave basis.

Small compressors 60/300 CFM
Photograph by permission Lucas, Birmingham

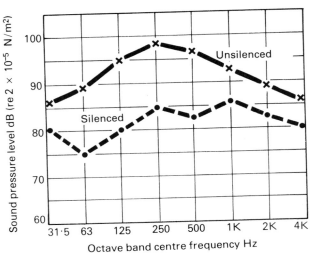

Large compressors 600/500 CFM
Photograph by permission Plant All Services, Birmingham

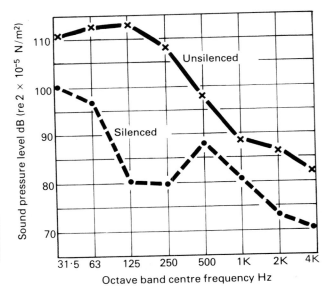

Claimed noise reduction 10dB(A) small compressors
16dB(A) large compressors

Cost (1981) £170–£355

* Blundell Acoustics Ltd, Coventry

Active control of gas turbine exhaust noise

The principle of Active Sound Control, or 'Antinoise', is that sound waves of equal and opposite amplitudes are deliberately superimposed in such a fashion that they cancel each other out, resulting in considerable sound reductions. This principle was first proposed many years ago, but only during the last decade has the development of active sound attenuation progressed from theoretical analysis, through laboratory experiments of increasing sophistication to a stage where practical application to certain large scale industrial sources has, at last, become a reality.

In this gas turbine exhaust silencer, a set of microphones, control electronics and loudspeakers are used to detect and generate the inverse sound field, which is then combined with the original sound generated by the machine, yielding a net reduction in level.

The 12m high and 3·3m diameter exhaust stack of the 11 MW gas turbine has placed around its exit 72·15 in bass loudspeakers driven by 12,1 kW amplifiers. The total cost of this hardware was considerably less than the estimated cost of conventional passive modifications necessary to achieve similar performance.

Installed system
Photographs by permission of British Gas Corporation and Topexpress Ltd

Speakers before installation

10 dB

Results achieved with gas turbine active attenuator
Power spectra: Attenuator off and on

Claimed noise reduction 10–12 dB in 31·5 Hz octave band

Design and installation by Topexpress Ltd, Cambridge in consultation with M. A. Swinbanks, Cambridge, funded by the National Research Development Corporation, London SW1, in collaboration with British Gas Corporation, Newcastle upon Tyne

Suspended noise absorbers

Over 1500 spatial noise absorbers, mounted in the roof of the maintenance and repair workshop of one of ICI's major divisions in the U.K. have considerably reduced workshop noise levels and improved the acoustic environment. High intermittent noise levels from fabrication and machining equipment, coupled with the disturbance from hammering and presses, made the working environment unacceptably noisy.

Installing the spatial noise absorbers – the standard product is a 900mm × 600mm × 50mm absorbent slab sealed in an acoustically transparent plastic foil bag – has reduced the reverberant noise by some 6dB(A) and softened the acoustic environment by reducing the measured reverberation time from 4 seconds to about 1 second. The absorbers were suspended easily at roof level with minimum disturbance to lighting and services. They are spaced at approximately one per square metre of plan area which is normal for most industrial applications.

The absorbers are suspended either from catenary wires spanning the factory or from metal 'T' section supports fixed to the underside of existing roof trusses.

Claimed noise reduction 6dB(A)
Approximate cost (1981) £6 to £10 (per absorber) 'installed' at 1 per sq metre of plan area and depending on quantity and site situation.

* Design and installation by ICI Acoustics, Welwyn Garden City, Herts

Suspended sound absorbers

There are two components of noise contributing to the total noise exposure of an operator in a workshop: direct noise due to an operator's own machine and immediately adjacent machines, and reverberant noise due to the general noise in the workshop, this latter reverberant noise level or general noise is a function of the acoustic characteristics of the workshop. Generally the more acoustically absorbent the workshop structure is – walls, ceiling etc., the lower is the reverberant noise level. One company by installing sound absorbing panels* has reduced reverberant noise levels and as a result has considerably improved working conditions in their auto lathe shop.

Some 800 sound absorbent panels each of mineral wool in a knitted nylon cover have been suspended from spiral springs just below the ceiling of the workshop. The panels have dimensions 900 × 600 × 75mm and are arranged in a regular parallel pattern at a 'density' of approximately 1 per square metre of plan area.

The workshop containing multispindle automatic lathes and rotary transfer machines measures 30·3m × 24·6m × 4·48m high. The ceiling is concave and of plastered brick, the floor is concrete and two of the four painted brick walls are 40% glazed. The subjective impression (unsolicited) of numerous employees was that the installation of the sound absorbers had made a significant difference to the noise environment, making conversation possible without voice strain.

Photograph by permission of Hawke Cable Glands Ltd, Ashton-under-Lyne

Claimed noise reduction 4–6dB(A) (reverberant noise only)
Cost (1981) £4830 or £6·5 per sq. m. (installed)

---

* Supplied and installed by Acoustic Control Systems, Stockport, Cheshire

Suspended absorbers

Suspended absorbers reduce noise by shortening reverberation times, hence reflected build up. Workers whose hazard comes only from the direct path of the machine they are operating would gain little benefit. However, particularly in automatic shops when workers move about receiving mainly reflected noise from a variety of distant sources, reducing the overall noise environment can give noticeable improvements.

Such a system was installed in the ceiling of a press shop using absorber sizes 900 × 600m of 50 and 100mm thickness reducing the excessive ambient noise by between 5 and 6dB(A) depending upon locations. It was chosen because otherwise all presses would need to be treated individually even those automatic presses normally remote.

Photograph by permission of Mallory Batteries Ltd, Crawley

Claimed noise reduction 5/6dB(A)
Cost (1981) £4–£6 (per absorber) installation suspension wires, hooks etc £3–£4

Design and installation by T Mat Engineering Ltd, Loughborough, Leics.

Partition wall

Non-supporting building walls are erected to simply divide one part of a building from another. Their effectiveness as batteries against noise transmission between areas varies tremendously. To comply with the Health and Safety at Work etc. Act 1974, the Offices, Shops and Railway Premises Act 1963 or just to improve a working environment for comfort or greater efficiency, the acoustic characteristics can be improved at little extra cost.

Many text books on noise control include basic principles for maximum transmission loss giving tables of figures, graphs and line diagrams but often based on theoretical partitions.

The wall* shown below is included in this series because it is a practical application of the incorporation of good acoustic principles and subsequent testing has shown the required improvement on conventional construction at specific frequencies has been achieved.

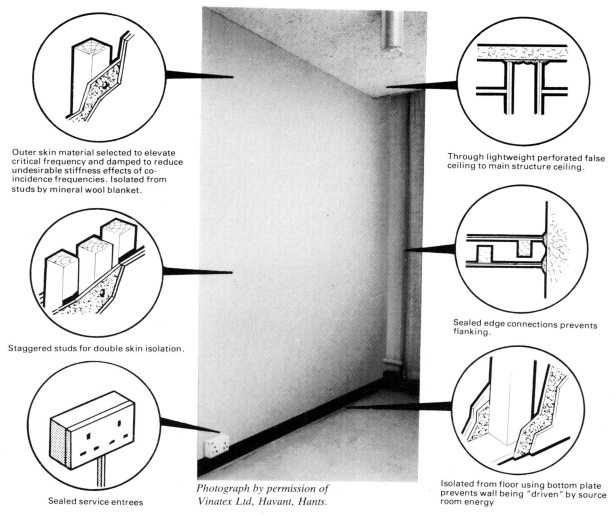

Outer skin material selected to elevate critical frequency and damped to reduce undesirable stiffness effects of co-incidence frequencies. Isolated from studs by mineral wool blanket.

Staggered studs for double skin isolation.

Sealed service entrees

Photograph by permission of Vinatex Ltd, Havant, Hants.

Through lightweight perforated false ceiling to main structure ceiling.

Sealed edge connections prevents flanking.

Isolated from floor using bottom plate prevents wall being "driven" by source room energy

Transmission loss

Octave band centre frequency Hz	63	125	250	500	1K	2K	4K	8K	Mid frequency average
SRI (dB) Wall shown above	26	35	37	42	50	54	54	57	44
Typical SRI (dB) conventional wall	20	20	17	27	33	34	42	48	26

Improved noise reduction 18dB (mid frequencies)
Estimated cost 10% additional over conventional partition.

* Designed by Ian Sharland Ltd, Winchester, Hants.

Acoustic screen reduces exposure of web press operators to noise

Noise has traditionally been accepted as a way of life for operators of web offset printing presses. It is now realised however that noise can cost money. Not only does it damage the hearing of operators but it may also limit production if a press becomes unworkably noisy at high speeds.

The work pattern of operators is such that the majority of their time is spent at the control console and inspection desk which is often sited conveniently close to the print take-off point from the folder unit and in front of the print stands, the two major noise sources which may contribute equally to the total noise at the control console area.

Whilst total enclosure of the print units can pose problems in some cases one firm has overcome these difficulties by installing an acoustic screen around its 4-unit press, separating it and the folder unit from the control and inspection area.

This has reduced noise levels at the press control console from 92dB(A)–81dB(A) and at the finished print checking position from 93dB(A)–83dB(A).

The screen comprises acoustic panels each with a large window of toughened glass which provide excellent vision of the machine. Quick and easy access through the screen to the press is via 4mm thick overlapping strips of transparent PVC.

Noise reduction 7–11dB(A)
Approximate Cost (1982) £17 500 (installed)

* Designed and installed by Bestobell Acoustics Ltd, Slough, Berkshire

Environmental haven for web press operators

Printing companies are becoming increasingly aware of the benefits of noise control measures. Not only does it reduce noise exposure of operators but inevitably improves working relationships. Reduction of noise at web presses, however, is no easy task and protection of operators by noise refuge is often rejected because of limited vision of the press and poor access to machine parts.

This has proved not to be the case at one printing company where refuges have very successfully reduced exposure and are fully accepted by the workforce who now 'wouldn't be without them'. Although the company who produce high quality magazine print had already reduced noise by enclosure of the folder units the workforce were still required to wear hearing protection throughout the shift.

A specialist acoustic company* was commissioned to further reduce noise levels and to provide good environmentally controlled working conditions for operators at (a) Baker Perkins G.16 five-unit web press (b) Two Harris Cotterell-Mariononi 1000 four-unit presses (c) Harris Cotterell-Mariononi 110 four-unit press.

'Quiet Room' enclosures were installed immediately adjacent to and alongside the lengths of the web presses to accommodate the operators working at the control consoles and those removing and stacking the finished print at the 'take-off'. The reel men working at the input ends of the presses have been provided with small individual noise refuges. The largest of the enclosures provided measures 25m × 7m × 2·3m high.

The enclosures are constructed of 100mm thick plastic-coated steel acoustic panels with large windows strategically positioned to provide maximum visibility of the critical print processes. Access to parts of the press for maintenance etc is by specially designed acoustic doors. A particularly beneficial feature of the enclosures is the provision of air conditioning systems to maintain optimum working conditions.

Automatic inking control minimises the time operators are required to be outside the noise refuge where noise levels may reach over 100dB(A) between the print units.

Outside quiet room – access via flexible curtains

Photographs and measurements by permission of Ben Johnson Limited, York

Inside quiet room – Noise levels 67–83dB(A)

Noise reduction 14–24dB(A)
Approximate Cost (1981) – Overall cost for all the noise reduction measures at the 4 presses listed above (including the air conditioning systems) – £51 000 fully installed and commissioned.

* Enclosure designed and installed by Tantalic Acoustical Engineering Ltd, Crayford, Kent

Quiet test house

Quality testing of products can require quiet test areas where noise from manufacturing processes do not interfere. However, efficiency often dictates that testing adjacent to production lines is preferable.

Test houses designed not only to accommodate the instruments and services required but also treated acoustically alleviate this problem.

The test house shown consists of four walls and a roof

of acoustic panels which were prefabricated and erected on site complete with observation windows and access door. To ensure the unit was isolated from surrounding noise sources antivibration strips were fitted under the wall panels and flexible connectors fitted to services passsing through the roof. Forced ventilation giving twenty air changes per hour was provided through a silenced duct located on the roof.

Photograph by permission of Phillips Domestic Appliances Ltd, Halifax

Claimed noise reduction 25dB(A)
Estimated cost (1981) £2500–£3000

Test House manufactured and supplied by T Mat Engineering Ltd, Loughborough, Leics.

Noise refuge

A large control room was required by a petfood manufacturer in a factory area which was frequently washed down. A modular cabin system* was chosen which combined with hygiene control the added advantage of protecting operators from excessive noise generated by can lines.

Extra large viewing windows are supplied on one side, and although all windows are single glazed, the glass reinforced plastic refuge achieves a noise reduction of 20dB to A-weighted sound to provide a working environment which is most acceptable to the occupants.

Photographs by permission of Pedigree Petfoods, Melton Mowbray

Noise reduction 20dB(A)
Cost (1979) £12 000 (included installation of special cable runs beneath the floor)

* Cabin by Glasdon Ltd, Blackpool, Lancs

Enclosure of fettling area solves combined noise and dust problems

Background

Always conscious of the need for environmental improvements the development team of a large steel foundry in Leeds set out to reduce the noise and dust problems in and around the fettling area.

The company installed an acoustic enclosure to separate the 12 operators engaged in the less noisy grinding and final finishing of small to medium range castings, from the noisier processes such as grinding of the large castings.

Noise control

An acoustic company* designed and installed the enclosure measuring some 17m × 11m × 3·6m high, constructed from modular acoustic panels. Before the enclosure was installed personal dosemeter measurements† on fettlers of small castings found exposures of 102–105dB(A) Leq. A repeat survey of the same occupational group at work inside the enclosure showed noise exposures of 93–99dB(A) Leq.

Dust control

Cold castings to be fettled are placed on a roller track and pushed along into the enclosure. Six men work on either side of the tracks at benches equipped with down draught extraction to remove dust as it is generated and discharge into water filter units away from the workshops. A roof mounted silenced ventilation unit makes up the enclosure air extracted by the dust removal process. Each operator has a heater directed onto his back to compensate cold air drawn in by the down draught benches.

Additional advantages

1. Low dust atmosphere inside the enclosure
2. Minimum interference: Since the enclosure is of modular construction should the area need to be extended/modified at a later stage this can be simply undertaken with the minimum of interference.

Photograph by permission of Catton and Co Ltd, Leeds

Claimed noise reduction   3–12dB(A) Leq (operators inside enclosure)
35–40dB(A) (to adjacent areas)
Cost (1981) £26 000 (outside shell of enclosure only. Total cost including heating, extraction benches, tracks etc approx £100 000)

* IAC Ltd, Staines, Middlesex
† SCRATA, Sheffield

Noise secure test cell

A large engine manufacturer called for the assistance of a specialist noise control company* to help in the design and installation of a 'noise secure' engine test cell to function within the main diesel engine assembly shop without adversely affecting the routine work operations.

The test cell required to effectively contain the high levels of noise generated by the diesel engines under test in order to provide essential protection to people working around the test cell.

The basic design of the cell was dictated by the fact that there was insufficient headroom for the overhead crane system to lift engines in and out for testing through the roof of the cell. Instead a unique front loading sliding acoustic canopy was designed. The canopy incorporates four inverted 'L' shaped sliding acoustic doors providing the front and roof to an existing brick and concrete test facility. The doors stack at both ends and can be opened in stages in each direction, providing convenient access through the front of the cell.

Each sliding door is constructed from a steel framework housing 100mm thick composite steel and acoustic infill panels and is mounted on wheels which run on tracks. Each door section has both wiper and deflective type seals to maintain the acoustic integrity of the cell with all doors closed. Movement of the doors is electrically controlled by an operator to provide the required access. The entire structure is designed to accommodate expansion caused by temperature rises inside the enclosure to 40°C during testing.

Photograph by permission of Mirrlees Blackstone (Stockport) Ltd, Stockport, Cheshire

Claimed noise reduction 30dB(A)
Estimated cost (1981) order of £32 000

* ICI Acoustics, Welwyn Garden City, Herts.

'Concertina topped' acoustic door

Background

Part of the factory layout of a large diesel engine manufacturer is such that the on-line mechanical test facility was originally open to an adjacent engine assembly area and also to a stores area at one side of the testing bay. As engine sizes have grown noise levels in the test facility have increased. The management, in order to provide greater protection from noise for both engine assembly and store personnel, called for the assistance of a specialist noise control company.*

Noise control

A two-fold 'design and build' solution was provided comprising a 'fixed' 100mm thick full height noise reducing wall between the stores and engine test bay and a separate 12 metre wide acoustic wall (shown above), incorporating part height fixed sections and the 'unique' concertina topped acoustic door assembly between the test bay and engine assembly areas. The unique door assembly itself consists of a large horizontal sliding door leaf housed within and having the same height as the fixed wall sections and a separate vertical lifting/folding concertina screen directly above and spanning the entire 12 metre width of the dividing wall below. Reaching a height of over 6 metres from floor level, the horizontal sliding acoustic door leaf, which is constructed from 100mm thick steel faced acoustic panels, is manually operated to slide across the face of the fixed wall to provide a clear opening of some 4·25 metres width.

A small single leaf door within the fixed wall construction provides personnel access when the main sliding door is closed. The whole of the large horizontal sliding door assembly is constructed to be stable with no fixed support at the head of the door opening, so that full height clearance is provided for the diesel engines to be swung through the opening beneath an existing overhead crane. Also, unrestricted travel of the overhead crane through a 4 metre high gap above the sliding door and fixed wall is accommodated by the vertical lifting/folding concertina design acoustic screen. When raised and folded this screen encroaches only 335mm down from roof truss level, but drops to fill the entire opening above the sliding door and fixed screen when in the closed position.

The concertina section itself is designed using a series of hinged folding lightweight panels, each incorporating flexible sound insulating fabric. Full acoustic integrity of the whole screen is maintained by carefully designed seals at all mating edges.

Door closed
Photographs by permission of APE Allen Ltd, Bedford

Door open

Claimed noise reduction 30dB(A) between test bay and engine assembly areas.
Approximate cost (1979) £34,000 for the front screen and door installation as shown in photograph.

* Design and installation by ICI Acoustics, Welwyn Garden City, Herts.

Reduced noise for web press operators

The pattern of work of the operators of web offset printing machines is such that they spend the majority of their time, while the machine is running, at the finished print take off point adjacent to the folder unit which is a major noise source. Also periodic visits are made to the aisles between the printing stands to check ink levels when noise levels up to 100dB(A) may be generated. To reduce operator noise exposure a printing company has provided a noise refuge/control room to enclose the work area adjacent to the folder unit.

For production reasons the company also fitted an automatic inking control system one secondary effect of which is to minimise the time operators are required to be outside the noise refuge. The reel man, the operator who works at the opposite reel end of the press, where noise levels are generally lower, has been provided with a small, individual noise refuge.

The machine to which the above relates is a new 5 unit, 950mm web M850 Harris-Cotterell-Mariononi offset litho printing machine. The output finished print enters the noise refuge through a short acoustic tunnel to Baldwin Countoveyors inside the refuge/control room.

Outside enclosure – at the folder, 95dB(A)

Inside enclosure – at the console – 64dB(A)

Inside enclosure – at the stacker – 75dB(A)
Photographs by permission of Carlisle Web Offset Printers Ltd, Carlisle

Sound level difference 20dB(A)
Cost (1980) £15 000 (installed)

* Enclosure by Tantalic Acoustical Eng Ltd, Crayford, Kent.

Simple noise screens for tile sanding

Persons stacking and checking clay tiles for the furnace had been exposed to hazardous noise levels of about 92dB(A) from compressed air sanding machines. The firm protected its operatives simply by screening them from the noise, as shown below.

The screens are made of two sheets of 10mm plywood mounted on a 75mm × 50mm timber frame with vermiculite infill, and they have small slots through which the tiles pass from the sanders. Although the screens are only 1·2m high, the seated operatives are afforded 7dB(A) protection.

Photograph by permission of Steetley Brick Ltd, Keele Works, Staffs.

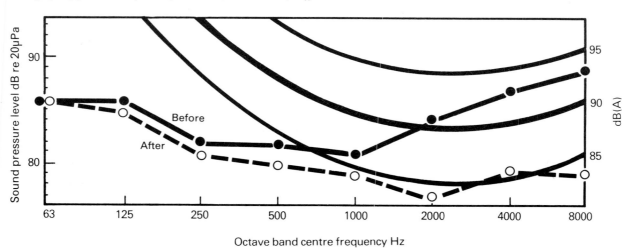

Noise levels before and after screening at a seated operative

Noise reduction 7dB(A)
Cost (1980) £30 per double screen

Panel treatment

For safety reasons, accommodation of complex controls or simply for aesthetic preference, the modern tendency is to box in machines with steel panels.

Far from reducing noise radiated from internal vibrating sources, direct fixing of these 'enclosures' can increase the total noise from the machine due to resonances, lower impedance and the increased area of vibrating surfaces which are more easily excited.

Very simple, low cost, treatments can be used to remove the adverse effects of these panels by addition of features which form the basis of acoustic close shielding.

The treatment shown is typical of the type added to existing machines. Noise Control Consultants assessing particular requirements may obtain better results by proper design of individual features, e.g.:

1 **Panel thickness:** selected to give optimum attenuation and away from coincidence frequencies
2 **Damping:** applied at the point or points of maximum vibration only to effect a saving in mass production runs
3 **Isolators:** selected for minimum vibration transmission
4 **Seals:** from materials to match contour while maintaining capacity
5 **Absorption:** different materials, possibly varying in density or thickness, to give maximum attenuation at desired frequencies
6 **Protection:** selection would depend on the application, possibly incorporated as skinning on absorbent lining
7 **Securing:** perforated steel may be preferred for ease of cleaning which could incorporate hole sizes to attenuate particular frequencies

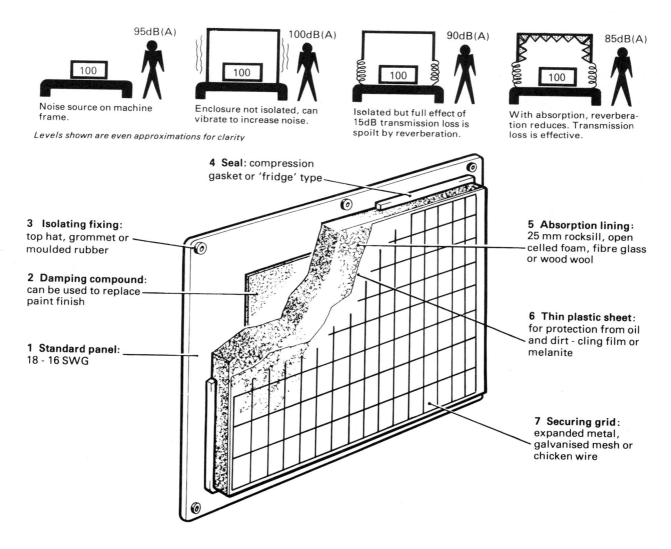

95dB(A)

Noise source on machine frame.

Levels shown are even approximations for clarity

100dB(A)

Enclosure not isolated, can vibrate to increase noise.

90dB(A)

Isolated but full effect of 15dB transmission loss is spoilt by reverberation.

85dB(A)

With absorption, reverberation reduces. Transmission loss is effective.

4 Seal: compression gasket or 'fridge' type

3 Isolating fixing: top hat, grommet or moulded rubber

2 Damping compound: can be used to replace paint finish

1 Standard panel: 18 - 16 SWG

5 Absorption lining: 25 mm rocksill, open celled foam, fibre glass or wood wool

6 Thin plastic sheet: for protection from oil and dirt - cling film or melanite

7 Securing grid: expanded metal, galvanised mesh or chicken wire

Noise reduction: 2–20dB(A)
Approximate cost (1981) £5 to £15 per square metre

Acoustic curtain reduces noise of vibratory bowlfeeder

Vibratory bowlfeeders are used widely in the bottling industry to provide a constant supply of bottle caps. The vibration ensures efficient feed of the caps to the capping machine, however the process is very noisy and can dominate the noise exposure of the capping machine operator.

As a trial a whisky bottling company fitted standard acoustic barrier curtaining material (lined with 25mm thickness of polyurethane foam) around the vibratory bowl and modified the anti-vibration mounting system so that the rubber mounts only made contact with the base plate at one point.

Tests showed that a noise reduction from 94dB(A) to 83dB(A) at 1m from the machine was obtained.

Bowlfeeder before treatment
Photographs by permission of John Walker & Son Ltd, Glasgow

Showing acoustic curtaining around the vibratory bowl

Noise reduction 11dB(A)
Approximate cost (1982) £30 per feed bowl

Enclosures for vibratory cleaning machines

Two large vibrators used for cleaning metal tools have been enclosed using panels made from ¼ inch marine ply sheets separated by 2 inch wooden supports. The enclosures* are readily dismountable: the sides are held in place using coach bolts, and the ends with brass screws. A hinged lid provides easy access, and a hinged door allows the vibrator to be emptied.

Showing access via lid and hinged door for emptying vibrator. (The small panel with a handle, next to the operator also slides out)
Photographs by permission Abingdon King Dick Ltd, Birmingham.

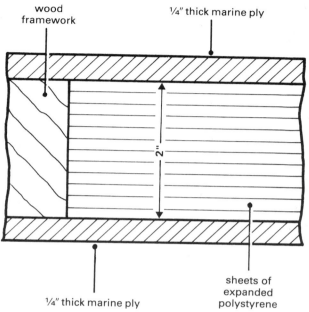

Sketch showing construction of enclosure panels.

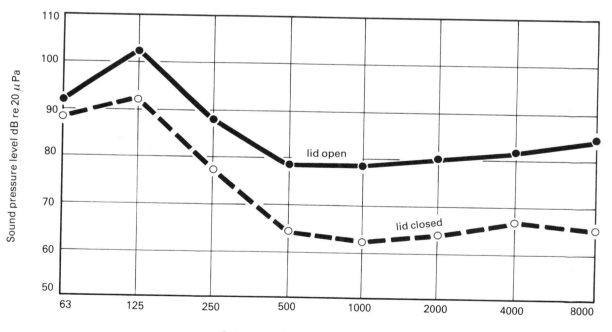

Noise reduction at operator's ear.

Noise reduction 16dB(A)
Cost (1979) £350 (built and installed by the company)

* Designed and installed by R T Atkins (Building Contractors) Ltd, Small Heath, Birmingham.

Noise reduction at wood hogging machine

A furniture manufacturer has achieved considerable reduction in noise by enclosure of their wood hogging machine which is used to break down scrap timber. The noise level beside the operator before enclosure was 118dB(A) sample Leq, after enclosure noise levels of less than 90dB(A) are achieved. The machine was enclosed by building onto an existing brick partition using hollow 150mm breeze blocks filled with sand, access doors were of substantial sound resisting construction.

Noise escape via the fixed chute was reduced by constructing an inner noise reducing chute to slide inside the existing metal chute of the machine. The inner chute was constructed of 15mm plywood in which slots were cut in the top to accept four pairs of battens which gripped strips of heavy flexible rubber which acted as a sound barrier. A hinged wooden lid provided further noise reduction.

View of innovative feed chute about to be slid inside existing machine chute. Note slots in top holding battens to grip rubber finger noise barriers.
Photograph by permission of PEL Ltd, Oldbury

Feed chute in use

Noise reduction 30dB(A)
Cost (1982) £300 (installed)

Company builds low cost enclosure

A company manufacturing ornamental brass pots and buckets has successfully reduced the noise generated by their planishing and spotting hammers, used to create decorative surfaces, by enclosing them in an enclosure built and installed by the company themselves resulting in a great saving in cost.

Four machines were housed at one end of a workshop along an outside wall and sectioned off from the rest of the shop by a half height partial partition of chipboard. Before enclosure noise levels of 110dB(A) sample Leq were measured at the planishing machine operator position, and levels of up to 95dB(A) were measured in the rest of the workshop.

The machines are now in a separate location from the main workshop and are enclosed as shown below. The enclosure is constructed of 12mm chipboard with a 2″ thick acoustic absorbent lining. Access to the hammer is through flexible 4mm thick transparent PVC strips. Enclosure has reduced noise levels to 93dB(A) sample Leq at the hammer operator position.

Photograph by permission of Derricourt Brassware, Birmingham.

Noise reduction 17dB(A)
Cost (1981) £120 (installed)

Engine test beds

A well known firm of diesel engine manufacturers has recently completed an extensive modernisation of their engine test bed facility – effective noise reduction was an important consideration at the design stage.

The new arrangement consists of an enclosure within a factory building, containing 8 separate test cells, arranged 4 each side of a central aisle. The walls are 200mm blockwork with substantial reinforced concrete roof and floor. The acoustic lining of each test cell is 50mm thickness of mineral wool covered with a polythene membrane and held by perforated steel sheet, fixed to timber battens. Heavy acoustic doors with a sound reduction rating of 38dB at 500Hz are fitted as shown. A large double glazed window is provided to each cell consisting of 6mm armour glass with a 150mm air space between panes. All engine performance monitoring is done from the central aisle.

'With a 210hp 6-cylinder turbo-charged engine running the noise level inside the test cell varied between 98 to 104dB(A). These levels were measured with the exhaust discharging to atmosphere prior to being extracted by the shop system. Measurements were taken adjacent to the engine and do not represent typical conditions when installed by the user. The level measured in the central aisle adjacent to the window was 66dB(A).'

View through a test cell towards the central aisle

Photographs by permission of R A Lister & Co Ltd, Dursley, Glos.

View of the central aisle between the test cells

Noise reduction 32–38dB(A)
Approximate cost £4 per sq feet of surface (acoustic absorbent lining to internal surfaces of test cell roof and walls)

Advantages of machine enclosure

Background

A firm manufacturing ball bearings uses eight large machines to grind rough balls to a smooth, spherical shape. Unenclosed they generated sound pressure levels greatly exceeding 100dB(A) as shown in the Table below.

Noise levels

Activity	Sound levels dB(A)	
	Before enclosure	After enclosure
Grinding new ⅞″ balls in Koehler machine with 3ft disc	114	88
After 20 minutes of grinding	103	84

Results supplied by PERA, Melton Mowbray, Leicestershire.

Noise control

Therefore the machines were enclosed with 18 gauge steel panels mounted on 2 inch square tube framework and 2 inch thick mineral wool internal lining for increased sound absorption within the enclosure.

Additional advantages

1. Production on the enclosed machines was increased by about 20%, due possibly to improved operator working environment.
2. Absenteeism in the area had been running at about 10%, much above the factory average, but decreased to normal levels following machine enclosure.

Photograph by permission of Fafnir Bearings Ltd, Wolverhampton.

Noise reduction 16–19dB(A)
Approximate cost (1977) £14 000

Enclosures by MM Acoustics, Melton Mowbray, Leicestershire

Access in cold header enclosure

Cold heading machines produce high impulsive noise levels chiefly as a result of the cold forging process itself, machine structure vibration and, in many cases, pneumatic ejection of the product. High levels of noise prompted one firm to install a sound reducing enclosure, shown behind an unenclosed machine in the photograph below.

As can be seen, the telescopic enclosure slides open to give excellent access to the machine. A separate, smaller door was later fitted to the enclosure to allow the setter to easily gain frequent access to the machine.

This overcame the setter's tendency to leave the enclosure open and hence lose the benefit of noise reduction.

Unenclosed and enclosed machines

Enclosed machine: showing telescopic enclosure sides
Photographs by permission of Fafnir Bearings Ltd, Wolverhampton

Noise reduction 12dB(A)
Approximate cost (1978) £2000

Enclosed machine: showing separate access door

Enclosure by MM Acoustics, Melton Mowbray, Leicestershire

Acoustic enclosure gives quick easy access

Cold heading machines generate high levels of repetative impactive noise. A New Zealand company has recently developed an acoustic enclosure of 19mm thick plywood skin, lined with a 50mm thick fibreglass sound absorptive lining to reduce the noise levels at their machines which unenclosed generated 95dB(A) at 1·5m distance. The main design criteria was to maintain quick and easy access, and this has been achieved with a hood which can be raised to expose the sides and top of the machine. Hydraulic cylinders are used to lift or lower the hood at the touch of a button. Ventilation is provided and the hood is interlocked to prevent operation in the open position.

Machine with hood lowered in operating position

Photograph by permission of Spurway Cooper Ltd, Auckland, NZ

Machine being set up with hood raised

Claimed noise reduction 10dB(A)
Approximate cost (1980) (NZ $3000)
(118/FI/108/1982)

Information supplied by Dept. of Labour, Wellington, New Zealand.

Enclosure of high speed punch press

Background

A large manufacturer of parts for the motor industry was proposing to purchase a new punch press to produce parts from heavy gauge spring steel strip. However it was realised that the noise levels which would be produced by the press, capable of operating at up to 600 strokes per minute, would be hazardous to the hearing of the operator and anyone else in the vicinity of the machine.

Noise control

An acoustic enclosure* was designed for the press and is shown below. The punch press can be fed with steel strip and controlled from outside the enclosure. Completed punched parts and waste are automatically discharged through acoustic ports, the parts being delivered by conveyor and chutes to stacking trays. Full interior lighting and double glazed windows in each wall ensure that any malfunction during running is quickly observed and the machine can be stopped and rapid entry obtained to set matters right. The several access panels allow simple routine inspection and maintenance, and also major overhauls or resetting of the tool, large replacement or substitution parts can be delivered right up to the machine by fork lift truck. When in operation, heat build-up within the enclosure is prevented by the integral fan ventilation, which takes in and discharges air via silencers into the roof space of the factory.

The enclosure occupied a minimum of shop floor space and was constructed of standard components from the Moduline range which included all the special features which were required namely several access panels, observation windows, silenced ventilators; loading and discharge ports etc.

Photograph by permission Automotive Products Ltd, Leamington Spa.

Claimed noise reduction 30dB(A) (operator position)
Cost (1982) £6000 (installed)

* IAC Ltd, Staines, Middlesex

Power press line enclosure

A company involved in the production of metal end caps for various types of containers were faced with the problem of excessive noise from four lines each of three power presses.

Noise levels ranging between 95dB(A) and 108dB(A) were measured at various operator positions on the line.

A specialist firm of enclosure manufacturers were invited to design and construct an acoustic enclosure having full length access doors, visual inspection panels, internal lighting and integral ventilation.

Unenclosed three press line 95–108dB(A)

Photographs by permission of Crown Cork Co Ltd, Southall, Middx.

Enclosure showing full length access doors 70–80dB(A)

Noise reduction 20–30dB(A)
Cost (1981) – £9700 (ventilation, lighting etc £2500)

Enclosure by Acoustic Enclosures Ltd, Colchester

A lightweight, flexible, power press enclosure

A company as part of an investigation into methods of reducing noise from power presses has successfully and economically enclosed a Bliss 30/3C double ram blanking and cupping press. The enclosure* (approximate dimensions 2m × 2·5m × 2·8m) consists of overlapping strips of 3mm thick transparent PVC suspended from a box section steel frame and has a roof of 18 SWG steel lined with 50mm of fabric faced glass fibre.

Noise level measurements were taken around the enclosed press during quiet background conditions with the press running at 90 strokes per minute blanking and cupping four blanks a minute from 75mm wide × 1·1mm thick mild steel strip. The results indicated noise levels of 83–86dB(A) measured outside the enclosure, 1m from the enclosure walls, and 98–99dB(A) measured inside the enclosure. Typical frequency spectra of these measurements are shown above.

PVC Enclosure around power press
Photographs by permission of Renold Power Transmission Ltd, Didsbury, Manchester

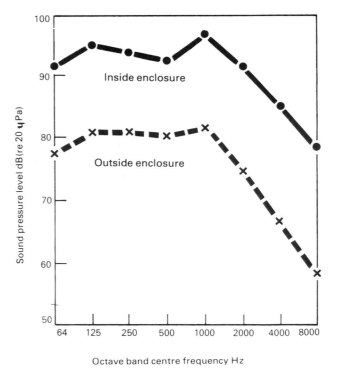

Noise reduction in octave band sound pressure levels

Noise reduction 12–16dB(A)
Cost (1981) £1100 (excluding installation estimated at £80–£150)

* Enclosure supplied by SRL Pollards Ltd, Bristol

Power press enclosure of flexible barrier material

A company who use a large number of power presses for the manufacture of motor parts have installed flexible acoustic enclosures around the largest and noisiest of the presses in their strip feed section.

The four sides and the roof of the enclosure* consist of curtains of acoustic barrier material lined with 25mm thickness of polyurethane foam. Each side is suspended from a tubular overhead rail along which the curtain may run so that quick easy access is provided to all press parts for setting up or maintenance. Velcro fasteners hold sections sealed when closed. All parts of the press were contained within the curtains which rested comfortably on the ground. A sturdy perspex window was incorporated in one panel for essential viewing of the tool area by the operator. To reduce

noise leakage from the entrance slits for the strip feed these have been sealed with overlapping fringes of acoustic barrier material.

Other noise reducing techniques which have been used successfully on power presses by this firm are:

a) the use of multi stage progression tools which perform several operations for one stroke of the press. At the last operation the products are pushed downwards and away eliminating the need for air ejection which is usually a major noise source.

b) introduction of a ½%–1% shear angle on tool cutting surfaces which also relieves the strain on the press machine.

Enclosure showing curtain opened for access

Enclosure closed

Showing several enclosed presses
Photographs by permission of Trico-Folberth Ltd, Brentford

Noise reduction 10dB(A)
Approximate cost (1972/73) £2000 (installed)
Estimated cost (1982) £8000

* Enclosure designed and installed by Bestobell Acoustics Ltd, Slough, Berkshire

Acoustic enclosure of a vibratory bowlfeeder

Bowlfeeders are often used in industry to feed small components such as: bottle caps, spools, bearing rollers, film cassettes, or, as in this case, brass electric terminal preforms to other processing or assembly machines. They operate by vibrating the small components which rest on the surface of the 'bowl' to move along a track. High noise levels are generated by the multiplicity of component impacts on the bowl surface and on each other. One company decided as part of a noise reduction programme to make noise enclosures for its bowlfeeders and the first of these built for a 10″ vibratory bowlfeeder is shown below.

The enclosure is constructed from 16swg mild steel lined with 28mm of polyurethane foam and is fitted with a 6mm thick cover of polycarbonate, hinged at one side and held against its seal by a magnetic catch.

The effect of the enclosure on the octave band sound pressure levels measured at 0·5m from the bowlfeeder casing and at a height of 1·5m is shown above.

Enclosure of a 10″ vibratory bowlfeeder

Photograph by permission of G H Scholes & Co Ltd, Wythenshawe, Manchester

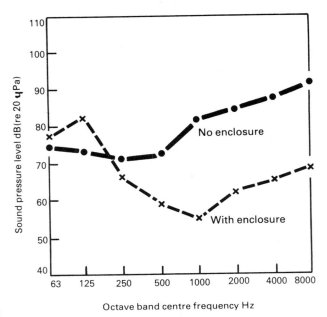

Effect of enclosure on octave band sound pressure levels

Noise reduction 23dB(A)
Cost £100

Enclosure of a wire straightening and cutting machine

It is generally found that automatic or semi automatic machinery more readily lends itself to acoustic enclosing. A New Zealand company involved in the manufacture of arc welding electrode rods has developed an enclosure for its fully automatic wire straightening and cutting machine.

The enclosure shown in the photographs has been designed with acoustically sealed hatches along its length which allow quick and easy access to all parts along the machine. The hatches have been provided with counter weights to ease opening. After enclosure the noise level in the area surrounding the machine was reduced from 96dB(A) to 84dB(A).

Photographs by permission of Weldwell (NZ) Ltd, Napier, New Zealand

Claimed noise reduction 12dB(A)
Approximate cost (1981) (NZ $3000)

Information supplied by Dept. of Labour, Wellington, New Zealand.

Enclosure of high speed rip saw

A furniture manufacturer has successfully reduced high noise levels generated from their multi-rip saw woodworking machine by enclosure. The machine uses 8 high speed rip saws to cut up 6ft lengths of hardboard for backing boards in domestic furniture and generated noise levels of up to 102dB(A). The acoustic enclosure* which measures 6m long by 2·5m wide is constructed of standard acoustic panels with mineral wool infill and has rubber mounted windows.

Dust is extracted by flexible ductings over each saw which is piped directly into the general factory extraction system. A separate extractor fan removes heat built up inside the enclosure.

Photograph by permission of Woodberry Bros and Haines Ltd, Highbridge, Somerset

Claimed noise reduction 22dB(A)
Cost (1979) £3500 (installed)

* Trubros Acoustics Ltd, Kegworth, Derby

Enclosures for wood draw-front machines

Two draw-front machines for drilling and grooving chipboard panels of self-assembly furniture exposed the operators to noise levels exceeding 100dB(A).

Enclosure of the machines has reduced the noise levels by almost 30dB(A). Each enclosure is made of 18g steel lined internally by 75mm mineral wool (100kg/m³ density) which is protected by 22g perforated metal. Access doors are provided at the front and rear and the viewing windows are double glazed. Both enclosures have been fitted with dust extraction, a sprinkler system and lighting.

The high noise reduction is achieved by having long, narrow sound reducing tunnels for feed and off-take of the chipboard panels (300mm at the feed and 600mm at the delivery). The company developed its own method of automatically feeding panels into the narrow tunnel using silenced pneumatic rams.

The two enclosures

Pneumatically operated feed system NB. silencers fitted

View showing narrow sound reducing delivery tunnel
Photographs by permission of Q A Furniture Ltd, Banbury, Oxon.

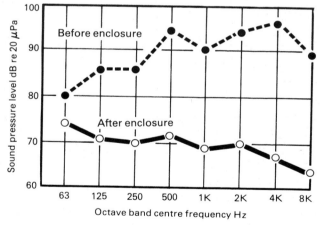

Figures before enclosure supplied by Alpha Acoustics Ltd
Noise levels measured 1 metre from delivery at operator head level

Noise Reduction 30dB(A)
Cost £3500 (1980) per enclosure including engineering alterations and installation.

Enclosures by: Alpha Acoustics Ltd, High Wycombe, Bucks

Enclosure for panel trim saw

Background

Panel trim saws consist of several independently driven circular saws mounted in a row above either a feed conveyor or traversing feed table. The saws are used to cut sheet material into smaller panels. The noise emitted depends not only on the characteristics of the saw and feed rate used but also on the nature, thickness and area of the panel being cut.

Noise control

An acoustic enclosure* has been developed for these machines which is based on 50mm thick acoustic panelling with a steel outer skin and perforated galvanised steel inner skin. Ready access for saw adjustment is available by full width counterweighted rising panels provided on the feed and take off sides. Windows are in 6mm clear glass. The enclosure is also available with a timber outer skin.

Noise levels

Machine: Wadkin WR Panel Saw of 120 in capacity fitted with 7 saws running at 3000 rev/m – Saw diameter 315mm

Material: 15mm thick chipboard with melamine laminate

Microphone position	Sound level dB(A)		Effective attenuation dB(A)
	Enclosed	Unenclosed	
Feed operator's ear	88	104	16
Take off operator's ear	84	100	
Inside the enclosure		108	

Photograph by permission of Hygena Ltd, Kirkby, Liverpool

Noise reduction 16dB(A)
Approximate cost (1981) £5000 (Steel outer skin)
£3500 (Timber outer skin)

Enclosure by Sound Solutions Ltd, Bradford, Yorks.

Thicknessing machine enclosure

As part of their investigations into woodworking machinery a college building department have developed a cheap simply erected acoustic enclosure for a thicknessing machine.

Made from ¾ in chipboard lined with acoustic foam it has proved to be effective and practical over months of use.

Enclosure removed, standing in the background

Enclosure fitted, note quick release clips

Approximate noise reduction 13dB(A) cutting softwood
Estimated cost (1981) £100

By permission of Glasgow College of Building & Printing.

Enclosure of semi automatic wood turning lathe

A company specialising in the manufacture of cricket bats were concerned with the noise exposure of the semi automatic wood turning lathe operator. A sound level of 102dB(A) had been recorded at the operators position.

An enclosure was constructed using 19mm chipboard panels mounted on a framework of 50 × 50mm timber. A number of the panels were made removable to give access to the tool and other areas. The inside of the en-closure is lined with 50mm thickness of sound absorptive material covered with a thin membrane of polythene sheet protected by wire mesh. A side door enables clearance of the wood chip and dust not removed by the extraction system fitted over the cutting area. The operator has access to the lathe by sliding open a double glazed counter-weighted window. This counterbalancing of the window gives quick and easy access.

Photograph by permission of Gray Nicholls Ltd, Robertsbridge

Noise reduction 20dB(A)
Cost (1981) £750

Enclosure of multi-head moulding machines

For some years the high levels of noise associated with the operation of multi-head planing and moulding woodworking machines have been of concern internationally to the manufacturers and informed owners of the machinery, and to organisations concerned with the health of employees.

Most manufacturers have attempted, with some success, to reduce noise at the machine design stage by experimentation with factors governing the generation of noise such as speed of rotation, the design of cutting tools etc. However such modifications at present are limited and certain manufacturers have opted for custom built enclosures, and these are available on many of the newest machines. Many companies are now building

their own enclosures for their woodworking machines particularly since the required materials are usually freely available to companies operating such machines. An enclosure is shown below which was built by a company in New Zealand for the 4 sider moulding machines which they were operating. It is constructed of 19mm chipboard lined with 50mm sound absorbent glass fibre matt retained with perforated sheeting. Particular attention was paid to reducing openings in the enclosure to a minimum, good door and window seals. The photograph below shows overlapping flexible fingers of heavy curtaining over the aperture of the discharge to keep noise escape to a minimum.

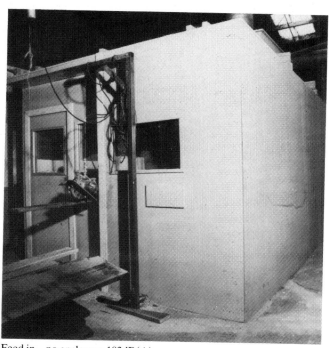

Feed in   no enclosure 103dB(A)
          with enclosure 88dB(A)

Photograph by permission Carter (Holt) Ltd, Hawkes Bay, New Zealand

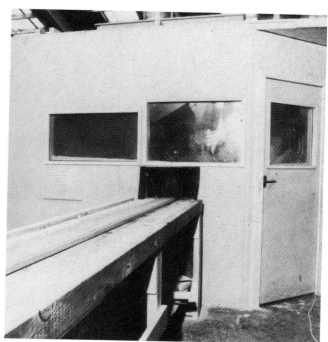

Discharge   no enclosure 105dB(A)
           with enclosure 84dB(A)

Claimed noise reduction   feed in   15dB(A)
discharge 21dB(A)
Approximate cost (1981) (NZ $3000)

Information supplied by Dept. of Labour, Wellington, New Zealand.

Open topped enclosure

The Woodworking Machines Regulations 1974, Regulation 44 requires that reasonably practicable measures to reduce noise must be taken when it is likely that an employed person is exposed to a continuous noise of 90dB(A) or its equivalent for more than 8 hours on any day.

A firm has successfully reduced the noise from their moulding machines by enclosing them in open topped enclosures with absorber panels suspended above the machines. The enclosures are open topped to allow for ventilation and sprinkler systems; absorber panels (baffles) are suspended directly over the moulder and improve ambient sound levels in the mill.

The enclosure is constructed of panels mounted on a wooden frame work, the panels are bolted onto the frame for easy dismantling. The panels are of ¾ in plywood lined with 2 in thick mineral wool and protected against dust with a thin polythene membrane. The windows are of a single panel of ¼ in wired glass set into a neoprene seal. The feed-in and take-off apertures were made as small as possible. At the take off a short acoustic tunnel had been fitted which reduced the noise to the operator, who stood approximately 16ft away, by a further 2dB(A). The tunnel incorporated simple adjustment for width and height of the aperture. The feed aperture was sealed with a flexible rubber barrier when the machine was free running, but when cutting, the aperture was open. A 20 in acoustic tunnel held temporarily in place at the aperture reduced noise to the operator by a further 3dB(A).

Absorber panels 4ft long by 2ft wide of ¼ in plywood base covered on either side with 1½ in mineral wool and protected by a light muslin cover were suspended above the machine spaced approximately 9 in apart and as low down to the machine as practicable.

Take-off

Photographs by permission Boulton & Paul (Lowestoft) Ltd, Lowestoft

Inside enclosure showing absorber panels

Noise reduction 12dB(A) (feed operator) 10dB(A) (take-off operator) both with acoustic tunnels
Approximate cost (1981) £3000 (installed by the company)

Close fitting moulding machine enclosure

Regulation 44 of the Woodworking Machines Regulations 1974 states that reasonably practicable measures to reduce noise must be taken when it is likely that an employed person is exposed to a continuous noise of 90dB(A) for eight hours or an equivalent noise exposure on any day.

A major noise problem for the timber industry is multi-cutter moulding machines which are capable of generating a serious hearing hazard, with typical noise levels falling within the range 95–110dB(A). A close fitting acoustic enclosure has therefore been developed* for these machines which is based upon a telescopic, sliding door system.

As part of a noise programme a large furniture manufacturing company purchased the enclosure shown above for a manually fed Weinig 17A+8 moulding machine. The enclosure consists of a fixed rear section through which all the machine services pass and for the front a contoured telescopic door system that is capable of being stored in the open position, outboard of the enclosure end panels, as shown above, thus giving total physical access along the length of the machine. Seals are employed between the sliding sections to provide a noise tight interface. Visual access is provided by single or double glazed windows incorporated into both the telescopic door system and the enclosure end panels. The windows can be easily replaced in case of accidental damage. Maintenance access to the drive motors is provided by either hinged or sliding doors in the rear of the enclosure.

Photograph by permission of Remploy Ltd, Swansea

Claimed noise reduction 14–25dB(A)
Cost (1981) £3100 (installed)

* Noise Control Centre, Melton Mowbray, Leics.

Proprietary timber moulding machine enclosure

A company specialising in the design and manufacture of modular building systems were faced with the problem of excessive noise from a six-cutter through-feed Wadkin moulding machine. Levels between 105 and 108dB(A) were recorded at both operator 'in-feed' and 'discharge' positions and permeated throughout the workshop.

The Woodworking Machines Regulations 1974 requires noise in excess of 90dB(A) to be reduced, so a specialist firm* were invited to design and construct a total acoustic enclosure which would not effect the machines productivity or functional requirements and maintain chip extraction facilities and service access.

The enclosure, illustrated below incorporates an access aperture for pressure adjustment of in-feed rollers, in-ternal lighting and emergency stop buttons. A reduction of 25dB(A) is claimed resulting in 80–82dB(A) at the operator's position.

Energy costs for maintaining an ambient working temperature of 13°C, required under Regulations, are high particularly due to extraction systems as used on these machines wasting heated air as they discharge under pressure to collectors outside the building. These enclosures can be provided with air through a duct fed directly from outside with consequent saving in heating costs. Extraction is still by the existing system which is in no way impaired.

A further bonus is the containment of dust and chip overspill which inevitably occurs. Work areas are cleaner and fire risk is reduced.

Photograph by permission of Austin Hall Building Systems Ltd, Huddersfield.

Claimed reduction 25dB(A)
Approximate cost (1980) £4000

* Noise Control Centre, Melton Mowbray, Leics.

A proprietary enclosure for moulding machines

As part of an extensive programme of noise reduction measures a company specialising in the large scale manufacture of kitchen and bedroom furniture purchased enclosures for 3 six cutter Wadkin FD moulding machines from a firm of acoustic enclosure suppliers.*

The machines were sited in a long, low (3m) and narrow (5·8m), room very close to one wall which formed one side of the enclosure. The other enclosure walls were of galvanised steel panels lined with a sound absorbent material protected by perforated steel sheet. The enclosure incorporated 2 double glazed windows and 3 single leaf doors. Feed and take-off apertures were fitted with non-overlapping heavy rubber fingers. Narrow rectangular ducts lined with sound absorbent material and each incorporating a 90° bend provided silenced air inlets for the chip and dust extraction system.

Feed roller adjustment from outside the enclosure was possible.

Noise levels

Noise reductions can be influenced by various factors including tools in use, feed rate, characteristics of the material being worked and the proportion of workpiece length projecting from the enclosure.

Machining process: Oak (40mm × 18mm) being planed, radiused and grooved at 10m/min

Measurement position	Sample Leq
Inside the enclosure	102dB(A)
1m distance from enclosure – feed-end	83dB(A)
1m distance from enclosure – take-off	84dB(A)

Feed-end

Photographs by permission of Hygena Ltd, Kirkby, Liverpool.

Take-off

Sound level difference 18–19dB(A)
Approximate cost (1981) £3000 or £70 per square metre (installed)

* Sound Solutions Ltd, Bradford, Yorks.

Compact enclosure for multicutters

As part of their noise reduction programme a kitchen furniture manufacturer decided to enclose two multi-cutter moulding machines.

The enclosures* which are similar to those now being fitted as original equipment by manufacturers of edge banders and double end tenoners, are constructed of 14swg steel panels lined with 60mm of foam acoustic absorbent material covered with a perforated facing and mounted on a box section steel frame.

Ready access for tool and adjustments is provided by three, top edge hinged doors supported, when open, by gas struts. Large, double glazed, 6mm polycarbonate windows provide a good view of the interior which is lit by internal strip lighting.

The panels at the rear of the enclosure open, or are removable, for maintenance access and the whole arrangement is very compact: Length 3·5m, Width 1·7m, Height 1·65m.

Dust extraction is provided by a tapered manifold which enters through one end wall and runs the length of the machine with flexible branch pipes to the cutter heads.

The noise levels shown were taken when moulding short Ramin door top rails 19½ in long × 1⅝ in × 1 in section at 11 metres/min.

Unenclosed machine
Photographs by permission of Thos Eastham & Son Ltd, Fleetwood

Enclosed machine showing hinged doors
Noise levels inside enclosure 102dB(A)
Noise levels outside enclosure 84dB(A)

Noise reduction 14dB(A) (insertion loss at feed position)
Cost (1982) £4500 (installed)

* Supplied by Indusvent Engineering (Northern) Ltd, Manchester

Multicutter open topped enclosure

In order to comply with insurance requirements and also to minimise fire risk a woodworking company used open topped acoustic enclosures supplemented by panel absorbers suspended over the machine to reduce noise at a six cutter moulding and planing machine.

The enclosure is of ½ in mahogany ply panels faced with 2–3 in of mineral wool held in place with expanded aluminium and then covered with hessian to keep out dust. Windows are of ¼ in wired glass, input and take-off apertures are sealed by ³⁄₁₆ in rubber fingers. Doors are sited to give access for cutter changing and removal of electric motors. Panel absorbers 4–5 in thickness made of mineral wool with a ½ in core of 3 ply wood and protected by an outer cover of hessian are suspended in banks above the machine spaced at about every 9 in. There is a gap between the bottom of the enclosure to the ground of about 6 in which facilitates sweeping dust and chippings from inside the enclosure. Electric wiring has been rearranged so that the machine can be stopped from a control outside the enclosure, but can be switched on only from the inside.

Feed in enclosed – 85dB(A)
Not enclosed – 94dB(A)
Photograph by permission Coronet Timber Co Ltd, London N1

Inside enclosure

Noise reduction 9dB(A)
Approximate cost (1979) £500 (installed by the company)

Band resaw enclosure

Enclosure of band resaws in premises processing imported timber is becoming standard practice, often users design and build enclosures with their own labour. These enclosures can be as effective as proprietary designs and are usually cheaper.

A company built an enclosure around their 54 in Stenner resaw which reduced noise levels from about 102dB(A) to 90dB(A) at the feed position. The enclosure incorporates a recess at the feed side so that no changes in the physical layout of the controls was necessary. The amount of space available around the machine allows pass back of the sawn material by bogey.

The enclosure is constructed of ½″ high density fibre board on 2 in × 2 in batons and is lined on the inside with 1 in thick rockwool which is protected from contamination by dust etc, by a membrane of polythene.

Take-off showing door for blade changing (twice per day)

Photographs by permission of Weir Constructions Ltd, Coatbridge

Feed showing recess for machinery controls

Estimated noise reduction 12dB(A)
Approximate cost (1982) £1000–£1500 (raw material and internal labour)

Band resaw enclosure of flexible acoustic material

A company employed an acoustics company* to enclose their Stenner band resaw. The enclosure is of flexible loaded PVC, of surface density about 5kg/m^2 lined with 1 in of polyurethane foam surface treated so that dust is easily brushed off. At feed and take-off positions there are sliding panels of clear acrylic sheets which are carried on runners and can be pushed aside for band changing etc as shown in the photographs below.

The enclosure is supported on a light alloy frame, the vertical sections carry a layer of nylon felt and the panels a velcro strip fastener. Access is achieved simply by ripping apart the two surfaces. At one corner of each clear acrylic sheet there is a similar arrangement. Noise levels quoted above were obtained during 'flatting' of 4 in × 3 in × 55ft long spruce timber. The noise level

measured under similar conditions at the feed before enclosure was about 101dB(A).

Attention has been paid to minimise open areas in the enclosure through which noise may escape, A pelmet of acoustic material has been fitted just above the opening panels to provide better acoustic sealing at this area. The area below the feed table and the delivery is enclosed, also part of the area below the table at the feed.

Operators are pleased with the enclosure indicating that it does not interfere with work as much as anticipated. They have also found that the enclosure has provided a noticeable improvement in the dust extraction at the machine and that both they and the surrounding workshop stay much cleaner during cutting times.

Feed-end sample Leq 92dB(A)

Take-off sample Leq 91dB(A)

Enclosure open

Photographs by permission of Sumacon Luralda Packaging Ltd, Kings Langley

Noise reduction 9–10dB(A)
Approximate cost (1978) £400 (installed)

* Bestobell Acoustics Ltd, Slough, Berks.

Demountable band resaw enclosure

A woodworking company designed and constructed completely demountable, all wood, enclosures for both their Robinson 36 in Type EFT and Stenner 48″ band resaws.

The basic assembly is of 4mm plywood on a 100mm stud framing with a 12mm insulation board lining and glass fibre infill. Double glazed windows of 6mm wired glass provide good visibility of the cutting zone and band tracking. Ready access for normal adjustments and maintenance is provided by both full height and smaller hinged panels. However basic machine operating controls have been brought out of the enclosure.

Noise levels

Machine & Stenner 48 in resaw
Machining Process & Softwood (75mm × 225mm × 5,100mm) being deep cut at 10m/min

Measurement Position	Sample Leq
Inside the enclosure (up to 4dB(A) may be due to reverberant field)	104dB(A)
Feed operator, 1m distance from enclosure	85dB(A)
Take off operator, 1m distance from enclosure	81dB(A)

Photographs by permission of Howarth Timber (Ashton) Ltd, Ashton-under-Lyne

Sound level difference 15–19dB(A)
Estimated cost (1981) £1,500 (installed including all erection and electrical costs)

PVC strip enclosure for band resaw

Background

Glasgow College of Building and Printing organise courses for safety officers. Part of the course deals with noise hazards and their control. As practical demonstrations, noise control techniques have been applied to a number of woodworking machines. A Robinson 48″ band resaw used for cutting timber for the furniture department and demonstration purposes is located in a rather confined basement resulting in high operator noise exposure and speech interference during demonstrations.

Action

To facilitate demonstrations and normal working an enclosure consisting of 100% over-lapping 4mm thick transparent PVC strips with three sides hung on sliding gear track was erected around the machine. A roof of chipboard lined with fibreglass and covered with pegboard was added to complete the enclosure.

Noise reduction 13dB(A)
Approximate material cost (1980) £800

Enclosure of large wire braiding machines

Following the systematic modification of a number of copper-nylon braiding machines with a view to reducing noise at source, a company decided to fit free standing acoustic enclosures to their larger braiding machines.

The B & F Carter 24 and 26 spindle machines, with horn gear centres ranging from 6 and 8 inches have been fitted with enclosures of 16swg steel lined with 2 inches of mineral wool retained by perforated galvanised steel sheet. Ready access is provided by interlocked doors on three sides of the enclosure and large, double glazed windows give a good view of the interior. The natural circulation of air for cooling purposes is achieved by incorporating a silenced intake low down on one side of the side walls and forming an outlet vent around the rotating drum which can be seen projecting through the roof of the enclosure.

Photograph by permission of BICC Ltd, Leigh, Lancs.

Noise measurements

Machine – 8 in Horn gear, 36-spindle heavy duty vertical slide plate wire braiding machine.

Microphone position – 1m from the machine at a height of 1·20m.

Condition	Sound Level dB(A)	Octave band sound pressure levels							Operating Condition
		Hz	125	250	500	1000	2000	4000	
Before enclosure	100		82	91	92	93	95	91	Braiding 5 ends 0·45mm GI wire. Diam. over braid 31·34mm
After enclosure	76		76	75	73	70	67	63	Braiding 6 ends 0·45mm GI wire. Diam. over braid 36·27mm

NB Background noise level 75 to 76dB(A)

Noise reduction 18–24dB(A)
Cost (1981) £3500 (Installed including provision of interlocks)

* Supplied by ACE Manchester

Noise reduction of braiding machines

The machines produce cross layered wire for electrical filaments and are not strictly speaking braiding machines. They are however similar to braiding machines in operation, though their mechanism is less complicated.

82 machines were housed in part of a large building in rows of about 12 machines each. With 58 machines running the ambient noise levels in the ranks varied between 98 to 100dB(A) depending on position.

On changing premises the firm enclosed the machines in groups of 6 as shown below. This comprises an acoustic enclosure of 18 gauge sheet steel with an inner lining of aluminium foil faced polyurethane foam either 12mm or 25mm thick. The foil is protection against oil and graphite spray from machine. Access is provided by sliding doors of similar construction.

Noise levels of enclosed machines gave readings of 85 to 86dB(A) at similar measuring positions. This being the shared noise from other processes in the building. Tests on one machine gave 94dB(A) with door open and 76dB(A) closed, in a quieter environment.

The cost of enclosure is understood to be about £425 for each group of 6 machines.

Noise reduction 14dB(A)
Estimated cost (1980) per machine £80.

Horizontal braiding machine enclosure

A company producing continuous nylon hose for high pressure installations has successfully reduced the noise in the braiding workshops by enclosing its horizontal machines used for braiding nylon reinforcement in the hoses.

The firm have progressively enclosed all their horizontal braiding machines in their braiding workshop. The enclosures were designed built and installed by the firm which made a considerable saving in cost and allowed modification and improvements to be made to the design each time a machine was enclosed. Such improvements included reducing the open areas in the enclosure through which noise could escape to a minimum, improving door seals, access to braiding head etc.

The original enclosure which was built around a 24 carrier horizontal braiding machine, is shown below.

Photograph by permission of Alenco Hilyn Ltd, Enfield.

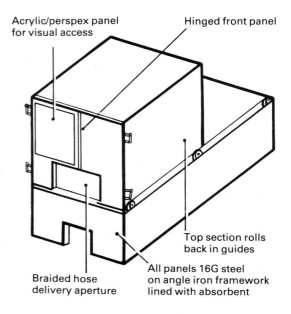

Acrylic/perspex panel for visual access

Hinged front panel

Top section rolls back in guides

All panels 16G steel on angle iron framework lined with absorbent

Braided hose delivery aperture

Noise reduction 9dB(A)
Cost (1974) £200

Enclosure for a confectionery demoulding process

The manufacture of some chocolate coated confectionery is by a shell moulding process. Metal tray moulds are coated with chocolate, filled with an appropriate sweet mixture and covered over with chocolate. After setting, the product is demoulded by inverting the tray and vibrating this against metal stops. This process produces high noise levels.

To reduce the noise a proprietary noise enclosure was provided,* designed for the particular process. The enclosure comprises a conventional sheet steel skin lined with a suitable acoustic absorbent. The sides and top of the enclosure form a split tunnel, mounted on rails to allow full access to the enclosed machinery for cleaning and maintenance purposes; the sliding parts are electrically interlocked at the closed position. Measurements taken before enclosure gave reading between 105–107dB(A), after enclosure reduced to 87dB(A).

Enclosure in closed position

Outfeed showing rail mounted split tunnel

Infeed showing removable top on acoustic tunnel

Claimed noise reduction 18–20dB(A)
Cost (1981) £6000 (installed)

* Noise Reduction Ltd, Eastleigh, Hants.
Information by courtesy of Cadbury Ltd, Keynsham, Bristol

Plastic granulator treatment

Granulation can lead to noise levels between 85 and 120dB(A) depending on the type of plastic and method processing.

A reduction in noise is often achieved by segregation or enclosure leaving only access to the feed throat. Frequently this throat remains untreated and in consequence, there is little reduction in operator exposure.

The treatment shown below incorporates modifications to the machine itself. One major feature is acoustic absorption and baffling in the throat which if incorporated into machines where segregation or enclosure has been used, would result in reduction at the operator position.

Vertical Feed Granulator, Typical Leq for Loading Cutting Cycle with ABS Plastic 101dB(A)

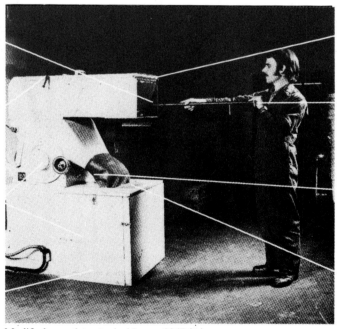

Close fitting 3 barriers in throat

Absorbent lining in hopper

Entire machine covered in mastic damping compound

Acoustic box contains product bin

Hopper squared off

Push rod horizontal feed

Heavy duty (commercial inner tube) at outfill

View window

Modified granulator typical Leq = 90dB(A)

Noise reduction 11dB(A)
Estimated cost (1980) £120

Plastics granulator enclosure

A manufacturing firm has successfully reduced the noise from plastics granulators by enclosure.

Treatment involved removing all sections of the throat and hopper above the granulating chamber and replacing by a new structure of similar internal dimensions though the base of the new hopper is 5 in longer than on the old machine. The walls of the new throat are of 12G steel where granules strike, but otherwise throughout the enclosure is of 22G steel lined with about 3½ in of sound absorption material faced with a perforated steel plate. A hinged steel barrier is suspended at the point where the hopper and throat join and a flap of flexible acoustic material ¼ in thick rests on the sloping front of the hopper. The granulation chamber is housed in a cube of acoustic panels as described above set on a steel frame. There is a generous distribution of doors for access to the drive motor and collection bin (here the powder was sucked out to a cyclone). Two air intake ducts are acoustically treated. The hopper and throat as a unit are hinged and can fall sideways to give access to the cutters. The system is electrically interlocked.

Before enclosure – grinding nylon 107dB(A)
Photographs by permission Lesney Products Ltd, London E9

After enclosure – grinding nylon 87dB(A)

Noise reduction 20dB(A)
Cost (1980) £700 (installed)

* Enclosure by Tantalic Acoustical Engineering Ltd, Crayford, Kent.

Paper corrugator noise enclosure

Most paper corrugating machines in use in the UK corrugate paper by passing it through a pair of inter-meshing, fluted rollers. Noise levels lie in the range 95–110dB(A) varying with size of the machine, fluted design and roller speed. The noise is characterised by the presence of strong pure tones between 125–1000 Hz.

Acoustic enclosures have been developed for these machines. The one shown below is a two storey enclosure providing access from a bridge catwalk as well as at ground level, supplied* for a Langston Masson single face corrugator.

Features of particular interest are: 1) the sliding panel covered aperture providing access for threading up, 2) the demountable panels facing the control panel which provide maintenance access, 3) the silenced ventilation system which minimises heat losses by ducting air to and from outside the building. This gives considerable financial benefit in terms of reduced heating costs.

NB As part of an extensive investment programme the above machine has now been phased out of production and been replaced by a 2·45m continuous run line.

Front view

Feed end

Inside enclosure

Photographs by permission Westbrook Packaging Ltd, Burscough near Ormskirk.

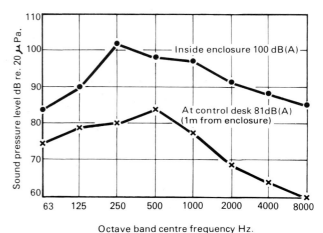

Noise levels inside and outside the enclosure

Sound level difference 19dB(A)
Cost (1982) £12–£15 000 (installed)

* Enclosures by Noise Reduction Limited, Eastleigh, Hants.

Foundry shake out enclosure

A foundry producing high quality non-ferrous castings, designed and constructed the enclosure system for the shake out area shown below.

The basic enclosure is constructed from unrendered concrete block walls 225mm thick to a height of 3·6m. The casting boxes and 'shaken' castings are craned in and out through a large opening in the enclosure frontal area 2.2m high by 2.75m wide onto or from the shake out grid within the enclosure. The open area is covered, in the front by 25mm thick overlapping plywood screens and on the top by a double thickness of plywood sheets infilled with mineral wool. Both screens are actuated simultaneously by a hydraulic cylinder through a simple cable and pulley system and are contained in substantial rolled steel channel section runners to effectively close off the open area.

Dust extraction is provided by a bag filter system drawing approximately 1m^3/s through a rectangular opening (300mm × 175mm) set high in a side wall of the enclosure. Sand from the shake out is collected in a skip within the enclosure. The whole arrangement is very robust and said to be trouble free. There was no visible signs of dust leakage from the enclosure. An additional benefit obtained by such an enclosure is substantially reduced air extraction costs.

Noise levels

Measurement position	Sampled Leq.
Inside enclosure	98–102dB(A)
1m distance in front of enclosure	88dB(A) (includes contributions from other foundry processes)
1m distance in front of enclosure	83dB(A) (shake out only operating)

Enclosure open

Enclosure c'osed

Sound level difference 10–15dB(A)
Estimated cost (1981) £1000–£1500 (excluding extraction equipment)

Information by courtesy of Meigh Castings Ltd, Cheltenham

Proprietary concrete block making machine enclosure

A company specialising in the manufacture of concrete pipes were concerned with the excessive noise emitted by their concrete block making machines. The sound generated by the machines varies according to the task being carried out. A typical work cycle included filling the mould, primary vibration, topping up of the mould and the final vibration. Sound levels of 109dB(A) to 112dB(A) were measured at the operator's position during the noisiest final vibration. There are three machines in the workshop and these high noise levels were excessive for everyone working in the vicinity.

The enclosure*, illustrated above, incorporates an aperture (not shown) for a conveyor to transport raw material to the top of the feed hopper. The hydraulically operated door shown at the front of the enclosure for removal of finished blocks is interlocked so that it is closed during the most noisy final vibration period. The sample continuous sound level measured at the operator's console during a number of work cycles was 90dB(A).

Photographs by permission of Milton Pipes Ltd, Sittingbourne.

Noise reduction 10–15dB(A)
Cost (1980) £5500 (installed)

* Enclosures by Sound Attenuators Ltd, East Gate, Colchester, Essex

Enclosure of paper bag making machine

A company who operate several paper bag making machines in New Zealand was concerned of the high noise levels in excess of 90dB(A) generated by their machines. They developed acoustic enclosures which, have resulted in the attenuation of the noise generated by the machines to levels of approximately 85dB(A). Good visibility of the machines is through transparent perspex panels. The enclosure lid may be lifted up quickly and easily with the aid of pneumatic cylinders for extensive access to machine parts.

Photograph by permission of U.E.B. Industries Ltd, Auckland, New Zealand.

Claimed noise reduction 15dB(A) (Beside machine)
                              12dB(A) (First packer position)
Approximate cost (1981) – (NZ $10 000)

Information supplied by Dept. of Labour, Wellington, New Zealand.

Industry index

Industry index

For quick reference to those case studies particularly relevant to certain industries these are indexed below. These should not be looked at in exclusion to other industry sections, noise reduction treatments are often general to all industries although a particular treatment may only be referenced against one industry.

HM Factory Inspectorate wish to thank

A.C.E. Manchester

A P E Allen Ltd, Bedford

Aarque Systems Ltd, Colnbrook, Bucks

Abingdon King Dick Ltd, Birmingham

Acoustic Control Systems, Stockport, Cheshire

Acoustic Enclosures Ltd, Colchester

Alenco Hylin Ltd, Enfield

R.T. Atkins (Building Contractors) Ltd, Small Heath, Birmingham

Alpha Acoustics Ltd, High Wycombe, Bucks

Anglian Building Products Ltd, Norwich

Austin Hall Building Systems Ltd, Huddersfield

Automotive Products, Leamington Spa

Avdel Fasteners Ltd, Welwyn Garden City, Herts

BICC Ltd, Leigh, Lancs

Babcock Construction Equipment Ltd, Gloucester

Barton Aluminium Foundries Ltd, Birmingham

Baynes (Reading) Ltd, Wheatley, Oxon

Berma Maschinenhandelsgesellschaft m.b.H.

Bestobell Acoustics Ltd, Slough, Berks

Biro Bic, London NW10

Blundell Acoustics Ltd, Coventry

Bonar Textiles Ltd, Dundee

Boulton & Paul (Lowestoft) Ltd, Lowestoft

British Gas Corporation

British Gas Corporation, Newcastle-upon-Tyne

British Railways Engineering Ltd, Shildon, Co. Durham

Distribution Unit, Building Research Establishment, Garston, Watford, Herts

Barton Aluminium Foundries Ltd, Birmingham

Cadbury Ltd, Keynsham, Bristol

Carlisle Web Offset Printers Ltd, Carlisle

Carter Holt (Central) Ltd, Napier, New Zealand

Catton and Co Ltd, Leeds

Dr F. Chatterton (Acoustics Group) Environmental Planning and Design Ltd, Basingstoke, Hants

Chester City Council, Chester

Compair Construction and Mining Ltd, Camborne, Cornwall

Cooper Roller Bearings Co Ltd, Kings Lynn, Norfolk

Coronet Timber Company Ltd, London N1

Crown Cork Co Ltd, Southall, Middx

Derricourt Brassware, Birmingham

Dolans Corrugated Containers Ltd, Bristol

Thos. Eastham & Son Ltd, Fleetwood

Epic Engineering Group, Slough, Berks

Environmental Planning & Design Ltd, Basingstoke, Hants

Fafnir Bearings Ltd, Wolverhampton

J. Fleming & Co Ltd, Aberdeen

R. Freeman, Agricultural Engineer, Craven Arms

GKN Floform Ltd, Welshpool

Gestetner Manufacturing Ltd, London N17

Glasdon Ltd, Blackpool, Lancs

Glasgow College of Building & Printing, Glasgow

Gray Nicholls Ltd, Robertsbridge, East Sussex

William Hackett Ltd, Cradley

John Haig & Co Ltd, Glasgow

Halton Borough Council, Chester

Hawke Cable Glands Ltd, Ashton-under-Lyne

A.G. Herbert, Wolfson Unit for Noise & Vibration Control, ISVR, Southampton

Joseph Horsfall, Halifax

Howarth Timber (Ashton) Ltd, Ashton-under-Lyne

Hygena Ltd, Kirkby, Liverpool

IAC Ltd, Staines, Middx

IAC Plastics, Burnley, Lancs

ICI Acoustics, Welwyn Garden City, Herts

Indusvent Engineering (Northern) Ltd, Manchester

Iver Plastics Ltd

Ben Johnson Ltd, York

Johnstone Pipes Ltd, Doselsey, Telford, Salop

Lamberton & Co Ltd, Coatbridge, Lanarks

Thomas William Lench Ltd, Warley, West Midlands

Lesney Products Ltd, London E9

R.A. Lister & Co Ltd, Dursley, Glos

London College of Printing, London EC1

Lucas, Birmingham

Lucas Industries Noise Centre, London W3

MM Acoustics, Melton Mowbray, Leics

Magnet Joinery Ltd, Keighley, West Yorkshire

Mallinson Denny Ltd, Cardiff

Mallory Batteries Ltd, Crawley

Meigh Castings Ltd, Cheltenham

Messer Griesham Ltd, Seaton Delavel, Tyne & Wear

MGD Graphic Systems Ltd, Preston, Lancs

Micor Ltd, Sheffield, Sth Yorks

Milton Pipes Ltd, Sittingbourne, Kent

Mirlees Blackstone (Stockport) Ltd, Stockport, Cheshire

Moniton Technic Ltd, Basingstoke, Hants

Monotype Corporation Ltd, Redhill, Surrey

National Research Development Corporation, London SW1

The New Zealand Department of Labour, Wellington, New Zealand

Noise Control Centre, Melton Mowbray, Leics

Noise Reduction Ltd, Eastleigh, Hants

The Northern Ireland Factory Inspectorate

North Western Newspaper Co Ltd, Blackburn

Oxford University Press, Printing Division, Oxford

PEL Ltd, Oldbury

PERA, Melton Mowbray, Leics

Palatial Ltd, London E3

Parker Foundry Ltd, Derby

FM Parkin (Sheffield) Ltd, Sheffield

Pedigree Petfoods, Melton Mowbray

Performance Plastics, Bacup, Lancs

Phillips Domestic Appliances Ltd, Halifax

Plant All Services, Birmingham

Q A Furniture Ltd, Banbury, Oxon

George Rankin & Co Ltd, Belfast

Remploy Ltd, Swansea

Renold Power Transmission Ltd, Didsbury, Manchester

Rolls-Royce, Bristol

R.H. Roseblade & Son, London NW10

Ryder Bros Ltd, Bolton

SP Civil Engineering Ltd, Brentwood, Essex

S.R.L. Ltd, Sudbury, Suffolk

SRL Pollards Ltd, Bristol

Salford University Industrial Centre Ltd, Salford

Sandell-Perkins Ltd, Aylesford, Kent

G H Scholes & Co Ltd, Wythenshawe, Manchester

Schrader Bellows Ltd, Bridgtown, Staffs

SCRATA, Sheffield

Ian Sharland Ltd, Winchester, Hants

WH Shaw & Son Ltd, Oldham, Greater Manchester

Singer and James, Ilford

Sound Attenuators Ltd, East Gate, Colchester, Essex

Sound Solutions Ltd, Bradford, Yorks

Spurway Cooper Ltd, Auckland, New Zealand

Steel Castings and Trade Research Association, Sheffield

Steetley Brick Ltd, Keele Works, Staffs

Stephen Laminates, Glenrothes, Fife

Stevens and Bullivant Ltd, Spring Hill, Birmingham

Sumacon Luralda Packaging Ltd, Kings Langley, Herts

M.A. Swinbanks, Cambridge

T Mat Engineering Ltd, Loughborough, Leics

Tantallic Acoustical Engineering Ltd, Crayford, Kent

3M Ltd Structural Products Group, Bracknell, Berks

Tilbury Plant Ltd, Maidstone, Kent

Topexpress Ltd, Cambridge

Trico Folberth Ltd, Brentford, Middx

Trubros Acoustics Ltd, Kegworth, Derby

Twyford Moors Aircraft and Engineering Ltd, Eastleigh, Hants

UEB Industries Ltd, Auckland, New Zealand

UK Provident Life Assurance, London

Vibrotechnique Ltd, Brighton, Sussex

Vinatex Ltd, Havant, Hants

W.I.R.A., Leeds

John Walker & Son Ltd, Glasgow

Weir Constructions Ltd, Coatbridge

Weldwell (N.Z.) Ltd, Napier, New Zealand

Westbrook Packaging Ltd, Burscough, near Ormskirk

Wolfson Unit for Noise & Vibration Control, ISVR, University of Southampton

JP Wood & Sons (Hatchery), Craven Arms

Woodberry Bros & Haines Ltd, Highbridge, Somerset

Wool Textile Economic Development Committee, Millbank, London

Printed in England for Her Majesty's Stationery Office by Eyre & Spottiswoode at Grosvenor Press, Portsmouth

Dd.715288 C.1000 10/83